Memoirs of the American Mathematical Society
Number 139

Erik Thomas

The Lebesgue-Nikodym Theorem for Vector Valued Radon Measures

Published by the
American Mathematical Society
Providence, Rhode Island
1974

ABSTRACT

Given a summable function $\vec{f}: T \to F$, defined on a measure space (T, β, ν) with values in a locally convex space F, it often happens that the integrals

$$\vec{\mu}(A) = \int_A \vec{f} d\nu$$

belong to a linear subspace $E \subset F$, itself endowed with a topology for which the inclusion map is continuous (e.g., Green's function, Fourier transform, "regularization"). When E is a separable or weakly sequentially complete Banach or Fréchet space we prove that the resulting set function $\vec{\mu}: \beta \to E$ is still countably additive, and thus a ν-continuous vector measure. Unless $\vec{f}$ is in fact an E-valued measurable function the measure $\vec{\mu}$ will not have a density in E with respect to ν or any other scalar measure. This article systematically studies E-valued measures with F-valued densities when $E \subsetneq F$, and gives answers to the following questions: Under what conditions does an arbitrary absolutely continuous E-valued measure have an F-valued density? Under what conditions is an F-valued density actually an E-valued density? How does the variation of the measure (with respect to some norm in E) affect the F-valued density? Known and new existence theorems are obtained by plunging the given space into a product of lines. We prove that a measure with values in a reflexive Banach space has a density if and only if it has locally σ-finite variation. The basic measure space is always a topological space T with the Borel σ-algebra β and a positive radon measure ν. We give an example of vector measure of bounded variation with a density in the given space E, and with a unique weak* continuous density in E'', going out of E.

INTRODUCTION. In analysis it happens frequently that, given a vector valued function $\vec{f}: T \to F$, which is summable for some positive measure ν, the integrals

$$\int_A \vec{f}\, d\nu$$

belong to some linear subspace $E \subset F$ itself endowed with a topology such that the inclusion map is continuous. In that case, if E is a separable or weakly sequentially complete Frechet space the map

$$A \to \vec{\mu}(A) = \int_A \vec{f}\, d\nu$$

is still countably additive in E (cf., §0) and hence is an E valued ν-continuous vector measure. However, in order that $\vec{\mu}$ have a density with respect to ν in the space E it is necessary (and not in general sufficient) that $\vec{f}(t) \in E$ ν-almost everywhere, which is not generally the case.

This explains why the classical Lebesgue-Radon-Nikodym theorem does not hold for vector valued measures (in §4 it is shown that it never holds in an infinite dimensional Banach space), and it motivates us to study the following situation: E and F are appropriate locally convex spaces with a given continuous injection $j : E \subsetneq F$. The map $\vec{\mu}: C(T) \to E$ is an E-valued Radon map or measure[1] such that $j \circ \vec{\mu}$ has a density with respect to ν (we shall describe this by saying that $\vec{\mu}$ has a density in F).

Received by the editors August 13, 1971.

[1] For definitions of Radon maps and measures see the end of this introduction.

Certain results show that such a density "going out of the space" can have properties similar to densities in some more conventional sense (cf., Theorem 2.9, Example 6.6).

Even when one wishes to establish the existence of a density for $\vec{\mu}$, in the space E it can be very convenient to construct a density $\vec{f}$ in an enlarged space and then to show that $\vec{f}$ takes its values essentially in E, and defines an E-valued density for $\vec{\mu}$ (cf., the proof of the Phillips-Grothendieck theorem in §3, and Proposition 6.10).

Finally it can happen that even when $\vec{\mu}$ possesses an E-valued density with respect to ν, there is a unique continuous density with values in F with almost all, but not all, the values in E, so that even when $\vec{\mu}$ has a density in E it can be of interest to consider densities "going out of the space."

The situation described above, where $\vec{\mu}$ is a Radon map with values in $E \hookrightarrow F$, having a density in F with respect to ν will be referred to as "the general situation." In many cases the spaces E and F will coincide as linear spaces, only the topologies being different. In that case the terms "density going out of the space" etc. become rather abusive and should be replaced by "density relatively to a weaker topology." We shall always clearly specify the topology under consideration.

In the present article we have taken a topological space T with its Borel subsets and an inner regular positive measure ν as the underlying measure space. This enables one to use the Lusin property as a definition of measurability for functions taking their values in arbitrary (not necessarily metrizable) locally convex spaces (e.g., the dual of a Banach space equipped with the weak* topology). However, it will be clear to the reader that all results involving only metrizable locally convex spaces can be written in the context of σ-finite or localizable measures on arbitrary σ-algebras.

The plan of this article is as follows: The introduction contains, after this section, some basic definitions and conventions, and a brief review of the relation between this article and the work of other authors.

§0 establishes generalizations of the Orlicz-Pettis theorem notably: If E is a separable or weakly Σ-complete Fréchet space and if $m: \mathcal{A} \to E$ is an additive set function defined on a σ-algebra such that $x' \circ m$ is σ-additive for a total set of continuous linear forms x', m is σ-additive.

§1 introduces the notion of vector valued summable function which will be used in this article.

§2 studies basic properties of Radon measures (or mappings) with densities in the given space, or in an enlarged space, and generalizes results of §0 from summable sequences to arbitrary summable functions.

§3 concerns universally measurable mappings and their relation to the present article; the theorem of Phillips-Grothendieck is established using results from §2.

§4 contains some of the main results of this article. Necessary and sufficient conditions on the injection $E \subsetneq F$ are established in order that any ν-continuous E-valued measure have a density in F.

§5 studies conditions under which an F-valued density is in fact an E-valued density, when $E \subsetneq F$.

§6 establishes circumstances under which the relation $\vec{f}(t) \in E$ a.e., for some F-valued density can be deduced from the variation of $\vec{\mu}$

with respect to some norm in E. Necessary and sufficient conditions for the existence of densities in reflexive or dual spaces are obtained.

§ 7 gives necessary and sufficient conditions for the existence of continuous (or weakly continuous) densities.

In this article the basic measure space is always a topological Hausdorff space, with its Borel σ-algebra (smallest σ-algebra containing the open sets) and an inner regular locally finite positive measure ν (positive Radon measure: cf., Bourbaki [2] Chapter IX, L. Schwartz [1]). However the search for densities of vector valued measures with respect to ν can be immediately reduced to the case where T is compact by breaking T up into compact pieces (cf., Bourbaki [2] Chapter IX, p. 18) and except in § 1 we have taken T to be compact.

The topological vector spaces used in this article will always be supposed to be locally convex quasi-complete[(1)] Hausdorff spaces, over $\mathbb{C}$ (resp. $\mathbb{R}$) and will be denoted by the letters E, F etc.

A vector valued <u>Radon mapping</u> defined on a compact space T is a continuous linear mapping $\vec{\mu}: C(T) \to E$, where $C(T)$ is the space of continuous complex (resp. real) valued functions equipped with the topology of uniform convergence. Such a Radon mapping will be called a <u>Radon measure</u> when it is weakly compact, or equivalently, when for every bounded Borel function g there exists $\vec{\mu}(g) \in E$ such that

$$\langle \vec{\mu}(g), x' \rangle = \int g\, d(x' \circ \vec{\mu})$$

---

[1]Quasi-complete: i.e., having its closed bounded sets complete. The dual of a Banach space with the weak* topology, the space of operators on a Banach space with the strong operator topology, are quasi-complete, but not in general complete.

for all continuous linear forms x' on E. The Radon mapping is then determined by the measure $A \to \vec{\mu}(A) = \vec{\mu}(\chi_A)$, which is regular in that

$$\vec{\mu}(A) = \lim_{K \subset A} \vec{\mu}(K)$$

where the limit is taken with respect to the filtering increasing compact subsets K of the Borel set A. Conversely a measure with this property determines a weakly compact Radon mapping, whence the terminology. In Bourbaki [2] Chapter 6 and Thomas [1] Radon mappings are called "mesures de Radon." In the latter article Radon measures are called "mesures de Radon prolongeables."

An arbitrary E-valued Radon mapping is a Radon measure if and only if E is weakly Σ-complete, i.e., satisfies the following condition: For every sequence $(x_n)_{n \in N}$ of points of E, such that $\Sigma_n |\langle x_n, x'\rangle| < +\infty$ for all $x' \in E'$, there exists $x \in E$ such that $\langle x, x'\rangle = \Sigma_n \langle x_n, x'\rangle$, (cf., Pełczynski [1], Bessaga [1], Thomas [1]). This condition is satisfied when E is reflcxive or weakly sequentially complete.

When E is a Banach space a Radon map $\vec{\mu} : C(T) \to E$ is said to be of bounded variation when there exists a positive Radon measure ν on T such that

$$|\vec{\mu}(\varphi)|_E \leq \int |\varphi| \, d\nu \qquad \text{for all} \quad \varphi \in C(T)$$

In this case $\vec{\mu}$ is a Radon measure and the smallest measure ν with the above property is the variation of $\vec{\mu}$ in the ordinary sense. When $\vec{\mu}$ is of bounded variation $\vec{\mu}$ can be extended uniquely to a continuous linear map from $L^1(\nu)$ to E. Conversely a continuous linear map from $L^1(\nu)$ to E defines a Radon measure of bounded variation by "restriction" to $C(T)$.

A function $f : T \to X$ from T to a topological space X is said to be ν-measurable if for every compact subset $K \subset T$ (T if T is compact) and every $\varepsilon > 0$ there is a compact $K' \subset K$, with $\nu(K - K') \leq \varepsilon$, such that the restriction $f|_{K'}$ is continuous (cf., Bourbaki [2] Chapter IV). When the topology of X is the coarsest for which mappings $u_i : X \to X_i$, $i \in I$, into topological spaces X_i are continuous, the continuity of f follows from the continuity of the functions $f_i = u_i \circ f$. However if I is not countable the ν-measurability of f does not follow from the ν-measurability of the f_i (because K' depends on i !). In spite of this many authors, probabily by analogy with the case of continuous functions, have termed the function f measurable when only the f_i are measurable, notably where X is a locally convex space E equipped with the weak topology. According to this usage a function is called weakly measurable when for every continuous linear form $x' \in E'$, the scalar function $\langle \vec{f}, x' \rangle$ is measurable. The chief inconvenience of this convention is the following: If f and g are functions with values on $X = \prod_{i \in I} X_i$ such that f_i and g_i are measurable for all i, it can happen that $f_i(t) = g_i(t)$ for almost all t, for every i, but that $f(t) \neq g(t)$ for all t. Thus if $\vec{\mu} = C[0,1] \to E'$ is a measure with values in the dual of a Banach space E, such that

$$|\vec{\mu}(\varphi)| \leq \int |\varphi| \, dt ,$$

there exist functions $\vec{f} : [0,1] \to E'$ such that for every $x \in E, \langle \vec{f}(\cdot), x \rangle$ is integrable and

$$\langle \vec{\mu}(\varphi), x\rangle = \int \langle \vec{f}(t), x\rangle \varphi(t)\, dt , \qquad \text{with } |\vec{f}(t)| \leq 1 ,$$

but, contrary to the assertion of Dunford and Schwartz ([1] VI. 8. 7), such functions are not essentially unique, (two may differ at every point), unless of course E is separable in which case such functions are in fact measurable relatively to the weak* topology $\sigma(E', E)$ (cf., Example 6. 11).

If we argue that the ultimate object of analysis is generally to prove that some "x" is equal to some "y", it is clear that the essential uniqueness of densities (and actual uniqueness of continuous densities) is a precious quality, and it actually simplifies the proof of the existence in certain cases, (through the use of Lemma 2. 4). We shall therefore avoid the above described convention on "weakly measurable" functions, but only use the general definition with the Lusin property relatively to the weak topology.

Relations with other work. The basic idea for the main result in §0, namely the use of the theorem of Banach Dieudonné is due to Grothendieck [2], who did not however state the relevant result correctly, since in an arbitrary Fréchet space a total family of linear forms does not have the Orlicz property (e.g., the coordinate forms in ℓ^∞ do not have the Orlicz property). The lemma proved by Grothendieck can be stated as follows: A family of bounded additive set functions defined on a σ-algebra $\mathscr{A}$, which is bounded at each set $A \in \mathscr{A}$, is uniformly bounded. We give another proof of this lemma, not involving the Stone-Kakutani representation space, but using an alternative inspired by the classical proof of the boundedness of a finite measure (cf., Dunford and Schwartz [1] III. 4. 4). All proofs involve a basic lemma by Phillips [1] p. 525.

The notion of summable function in §1 is related to Pettis' definition of summable functions, but different in that measurability is required. The absolutely summable functions, with values in a Banach space, are the same as the essentially integrable functions in the sense of Bourbaki [2] Chapter 5, themselves related to the functions which are integrable in the sense of Bochner.

Densities "going out of the space" as introduced in §2 have been used in the context of spectral measures as "eigenvectors going out of the space," and have been studied by numerous mathematicians and as "generalized eigenvectors." (Cf., Gel'fand and Shilov [1] Vol. 3).

The characterization of absolutely summing operators due to A. Pietsch [1] 2-3-3 is the basis of the main results in §4.

The idea of applying the closed graph theorem to the space of all measurable functions is due to L. Schwartz [2] who applied it in a different context. A further relation with Schwartz's work on Radonifying operators [2] is as follows: L. Schwartz calls Nikodymising a linear map $u : E \to F$ (E, F Banach spaces) which is such that for any weakly ν-continuous Radon map $\vec{\mu} : C(K) \to E$, $u \circ \vec{\mu}$ has an absolutely summable density with respect to ν. By applying this to the case of an atomic measure it immediately appears that u must be absolutely summing. In our Theorem 4.7 however we only ask for a summable density, but it still turns out that u must be absolutely summing, and consequently that the density must be absolutely summable. Thus it appears a posteriori that the two types of Nikodymising mappings (according to whether one asks for summable or absolutely summable densities) are in fact the same, at least in Banach spaces. To avoid confusion we have not used the term Nikodymising. The result in

Corollary 4.2 and several particular cases in 4.1 had previously been proved by L. Schwartz [2], [3].

The use of measures with σ-bounded variation bears a relation to Dieudonné's article Dieudonné [1].

§0. The Orlicz-Pettis property. Let E be a locally convex Hausdorff space. A family $(x_i)_{i\in I}$ of E is said to be summable, with sum s, when

$$s = \lim_K \sum_{i\in K} x_i ,$$

the limit being taken with respect to the filtering increasing finite sets $K\subset I$ (for any zero neighborhood V there exists a finite set K_0 such that for $K\supset K_0, s - s_K \in V$, where $s_K = \Sigma_{i\in K} x_i$).

When E is quasi-complete this is equivalent to Cauchy's condition: For any zero neighborhood V there is a finite set K_0 such that $s_K \in V$ whenever $K \cap K_0 = \emptyset$. In particular any subfamily of a summable family is summable.

Let E be a quasi-complete locally convex Hausdorff space; let H be a subset of the dual E' of E.

0.1 PROPOSITION. The following conditions on H are equivalent:

1) A family $(x_i)_{i\in I}$ of elements in E is summable if

a)
$$\Sigma_i |\langle x_i , x'\rangle| < +\infty \quad \text{for all} \quad x'\in H ,$$

and

b) For every set $A\subset I$ there exists $s_A \in E$ such that

$$\langle s_A , x'\rangle = \sum_{i\in A} \langle x_i , x'\rangle \quad \text{for all } x'\in H$$

2) The same condition for countable families.

3) Any set function $m: \mathcal{A} \to E$, defined on a σ-algebra $\mathcal{A}$ such that $x' \circ m$ is σ-additive for all $x' \in H$ is σ-additive.

Proof. 1) and 2) are equivalent since by Cauchy's criterium a family is summable if all its countable subfamilies are summable. 2) implies 3) by taking $x_i = m(A_i)$ where $(A_i)_{i \in \mathbb{N}}$ is a sequence of disjoint sets belonging to $\mathcal{A}$. Finally 3) implies 2) by taking $\mathcal{A} = \mathcal{P}(I)$ the set of subsets of I.

The theorem due to Orlicz, Banach, Pettis, asserts that these conditions are fulfilled when $H = E'$.

DEFINITION. When the conditions of the above proposition are satisfied the set H, or the topology $\sigma(E, H)$ induced by H, will be said to have the Orlicz property.[1]

0.2 THEOREM. Let E be a separable Fréchet space.[2] Let H be a total subset of E' (i.e., a subset such that $\langle x, x' \rangle = 0$ for all $x' \in H$ implies $x = 0$). Then H possesses the Orlicz property.

Proof. Let $\mathcal{A}$ be a σ-algebra (e.g., the set of subsets of $\mathbb{N}$) and let $m: \mathcal{A} \to E$ be a set function which is σ-additive relatively to the topology $\sigma(E, H)$. Let G be the subspace of E' consisting of all the $x' \in E'$ such

[1] More generally let $\mathcal{T}$ be a locally convex topology on E weaker than the given topology. We shall say that $\mathcal{T}$ possesses the Orlicz property when any set function $m: \mathcal{A} \to E$ which is σ-additive for $\mathcal{T}$ is σ-additive for the given topology. If H is the set of linear forms continuous with respect to $\mathcal{T}$, it follows from Orlicz theorem that $\mathcal{T}$ possesses the Orlicz property if and only if $\sigma(E, H)$ does.

[2] That is, a metrizable complete locally convex space.

that $x' \circ m$ is σ-additive. Then since G contains H, G is dense for the weak*topology $\sigma(E', E)$. By the Hahn-Vitali-Saks theorem G is sequentially closed for the weak*topology. It follows from the theorem of Banach-Dieudonné that G is closed for the weak star topology, hence that $G = E'$, whence m is σ-additive by the Orlicz theorem.

The above theorem is not as it stands a generalization of Orlicz theorem, even within the framework of Fréchet spaces. Let E be a locally convex space, and let $\mathscr{T}$ be a locally convex topology on E which is weaker than the given topology.

DEFINITION. We shall call E <u>essentially separable</u> with respect to $\mathscr{T}$ when any countable subset of E is contained in a linear subspace of E which is separable with respect to the given topology, and closed with respect to $\mathscr{T}$.

EXAMPLES. E is essentially separable with respect to $\sigma(E, E')$; if E is separable E is essentially separable with respect to any topology $\mathscr{T}$. If T is compact the space C(T) is essentially separable with respect to the topology of pointwise convergence. If I is an arbitrary set, the space $\ell^p(I)$, $1 \le p < +\infty$, is essentially separable with respect to the topology induced by the coordinates $x \to x_i$. The space $\ell^\infty(\mathbb{N})$ is not essentially separable with respect to the weak*topology $\sigma(\ell^\infty, \ell^1)$.[1]

0.3 THEOREM. <u>Let</u> E <u>be a Fréchet space and let</u> H <u>be a total subset of</u> E' <u>such that</u> E <u>is essentially separable with respect to</u> $\sigma(E, H)$. <u>Then</u> H <u>possesses the Orlicz property</u>.

---

[1] For details concerning these examples, cf., Thomas [1] Appendix II.

Proof. Since it is sufficient to verify condition 2) of Proposition 0.1 this follows immediately from Theorem 0.2.

Even without separability condition the Orlicz property is still valid for arbitrary locally convex Hausdorff topologies on E provided for instance E is a weakly sequentially complete Fréchet space. More generally:

0.4 THEOREM. Let E be a weakly Σ-complete Fréchet space, and let H be a total subset of E'. Then H possesses the Orlicz property.

Proof. Recall that E is weakly Σ-complete if the relations

$$\Sigma_n |\langle x_n, x'\rangle| < +\infty \quad \text{for all} \quad x' \in E'$$

imply that there exists $x \in E$ with $\langle x, x'\rangle = \Sigma_n \langle x_n, x'\rangle$ for all $x' \in E'$. It follows from Orlicz theorem that in this case a sequence $(x_n)_{n \in \mathbb{N}}$ is summable if and only if $\Sigma_n |\langle x_n, x'\rangle| < +\infty$ for all $x' \in E$, or what amounts to the same thing, if and only if the finite sums $s_K = \Sigma_{n \in K} x_n$ form a bounded subset in E. Let $A \to s_A$ be an additive set function defined on $\mathcal{P}(N)$ such that

$$\langle s_A, x'\rangle = \sum_{n \in A} \langle x_n, x'\rangle$$

for all $x' \in H$, in other terms, let $(x_n)_{n \in N}$ be summable, as well as all the subfamilies, relatively to the topology $\sigma(E, H)$. In the following Proposition we shall show that the s_A, and therefore the $s_K = \Sigma_{n \in K} x_n$ form a bounded set in E, and it will follow that $(x_n)_{n \in \mathbb{N}}$ is summable.

0.5 PROPOSITION (Grothendieck). *Let* E *be a Fréchet space and let* $m : \mathcal{A} \to E$ *be an additive set function defined on an* σ*-algebra. Suppose there is a total set of* $x' \in E'$, *such that* $x' \circ m$ *is bounded on* $\mathcal{A}$. *Then* m *is bounded.*

Proof. We shall use the following lemma:

0.6 LEMMA. *Let* Λ *be a subset of the space* $ba(T, \mathcal{A})$ *of bounded additive set functions defined on a* σ*-algebra* $\mathcal{A}$ *of subsets of a set* T. *Suppose*

$$\sup_{\mu \in \Lambda} |\mu(A)| < +\infty \quad \text{for all} \quad A \in \mathcal{A}.$$

Then

$$\sup_{\mu \in \Lambda} \sup_{\mu \in \mathcal{A}} |\mu(A)| < +\infty .$$

We admit this lemma for the moment and proceed with the proof: let G be the set of $x' \in E'$ such that $\sup_A |\langle m(A), x' \rangle| < +\infty$. By hypothesis G is a total subset, hence a $\sigma(E', E)$ dense subspace. If we show that $G = E'$ it will follow from the theorem of Mackey-Arens that m is bounded. In order to show that G is equal to E' it suffices to show that G is closed and by the theorem of Banach-Dieudonné it suffices to show that $G \cap B$ is closed for every $\sigma(E', E)$ closed and bounded subset B of E'. Since B is bounded

$$\sup_{x' \in G \cap B} |\langle m(A), x' \rangle| < +\infty \quad \text{for all} \quad A \in \mathcal{A}.$$

Hence by the previous lemma there is a constant M independent of A such that $|\langle m(A), x'\rangle| \leq M$ for all $x' \in G \cap B$. Then if x' belongs to the $\sigma(E', E)$ - closure of $G \cap B$ we still have $|\langle m(A), x'\rangle| \leq M$ or for all $A \in \mathscr{A}$, whence $x' \in G$, and B being closed $x' \in G \cap B$, proving that $G \cap B$ is closed. This completes the proof.

It follows that m may be extended to a continuous mapping from the set of bounded, $\mathscr{A}$-measurable function to E.[1]

Proof of Lemma 0. 6. We have successively:

1. Let $(\beta n)_{n \in \mathbb{N}}$ be a sequence of bounded additive scalar set functions on $\mathscr{P}(\mathbb{N})$ such that

$$\lim_{n \to \infty} \beta_n(\tau) = 0 \qquad \text{for every} \qquad \tau \subset \mathbb{N}.$$

Then

$$\lim_{n \to \infty} \sum_{i \in \mathbb{N}} |\beta_n(\{i\})| = 0 .$$

This is a lemma due to Phillips [1] p. 525.

2. Let $(\mu_n)_{n \in \mathbb{N}}$ be a sequence of bounded additive set functions defined on a σ-algebra $\mathscr{A}$ such that

$$\lim_{n \to \infty} \mu_n(A) = 0 \qquad \text{for all} \qquad A \in \mathscr{A} .$$

Then for every sequence $(A_i)_{i \in \mathbb{N}}$ of disjoint sets belonging to $\mathscr{A}$ we have

---

[1] The above proposition and its proof, are due to Grothendieck [1] (p. 217, French Edition, São Paulo 1964).

$$\lim_{n\to\infty} \sum_i |\mu_n(A_i)| = 0 .$$

It suffices to apply Phillips' lemma with

$$\beta_n(\tau) = \mu_n \left(\bigcup_{i\in\tau} A_i \right) .$$

3. Let Λ be a set of bounded additive set functions defined on $\mathscr{A}$, such that

$$\sup_{\mu\in\Lambda} |\mu(A)| < +\infty \quad \text{for every} \quad A \in \mathscr{A} .$$

Then for every sequence $(A_i)_{i\in\mathbb{N}}$ of disjoint sets belonging to $\mathscr{A}$,

$$\sup_{\mu\in\Lambda} \sum_i |\mu(A_i)| < +\infty .$$

It obviously suffices to prove this when Λ is countable. Suppose $\Lambda = \{\mu_n\}_{n\in\mathbb{N}}$. Then for every sequence of scalars $(c_n)_{n\in\mathbb{N}}$ tending to zero

$$\lim_{n\to\infty} c_n \mu_n(A) = 0 \quad \text{for all} \quad A .$$

Therefore by the previous result

$$\sum_i |c_n \mu_n(A_i)| = |c_n| \sum_i |\mu_n(A_i)|$$

tends to zero as n tends to infinity. The sequence $(c_n)_{n\in\mathbb{N}}$ being arbitrary it follows that

$$\sup_n \sum_i |\mu_n(A_i)| < +\infty .$$

Finally we show that if

$$\sup_{\mu \in \Lambda} |\mu(A)| < +\infty \qquad \text{for all} \qquad A \in \mathscr{A}$$

it follows that

$$\sup_{\mu \in \Lambda} \sup_{A \in \mathscr{A}} |\mu(A)| < +\infty .$$

To prove this we shall say that a set $B \in \mathscr{A}$ is bounded when

$$\sup_{\mu \in \Lambda} \sup_{A \subset B} |\mu(A)| < +\infty .$$

The union of two (disjoint) bounded sets is obviously bounded, and every set (belonging to $\mathscr{A}$) contained in a bounded set is bounded. The empty set is bounded and the object is to show that the whole space is bounded. Suppose the contrary, so that there are sets which are not bounded which we will call unbounded. Then we have either

a) For every unbounded set $F \in \mathscr{A}$ and every $M > 0$ there is an unbounded $A \subset F$ such that

$$\sup_{\mu \in \Lambda} |\mu(A)| > M$$

or

b) There is an unbounded set F and a number $M > 0$ such that for every unbounded $A \subset F$, we have

$$\sup_{\mu \in \Lambda} |\mu(A)| \leq M .$$

Suppose we are in case a). Then we can select a sequence F_n, and $\mu_n \in \Lambda$, with $F_n \supset F_{n+1}$, F_n unbounded, and $|\mu_n(F_n)| \geq n$. Let $A_n = F_n - F_{n+1}$ for $n \geq 1$. The sets A_n are disjoint and

$$F_1 = F_n + \sum_{i=1}^{n-1} A_i$$

(Σ and $+$ indicate disjoint union). Hence

$$\mu_n(F_1) = \mu_n(F_n) + \sum_{i=1}^{n-1} \mu_n(A_i) \quad .$$

Now by the last result

$$\left| \sum_{i=1}^{n-1} \mu_n(A_i) \right| \leq \sum_{i \geq 1} \left| \mu_n(A_i) \right|$$

is bounded independently from n, and $\mu_n(F_n)$ tends to infinity. Therefore $\mu_n(F_1)$ tends to infinity contrary to the hypotheses since

$$\sup_{\mu \in \Lambda} |\mu(F_1)| < +\infty \quad .$$

Suppose now that we are in case b). Let F and M be as specified. F is unbounded, so that

$$\sup_{\substack{A \subset F \\ \mu \in \Lambda}} |\mu(A)| = +\infty \quad .$$

Thus there is $\mu_1 \in \Lambda$ and $A_1 \subset F$ such that $|\mu_1(A_1)| \geq M + 1$. Then A_1 is bounded (for if not we would have $|\mu(A_1)| \leq M$ for all μ).

Since $F = A_1 \cup (F - A_1)$, $F - A_1$ is unbounded, and there exists $A_2 \subset F - A_1$ and $\mu_2 \in \Lambda$ such that $|\mu_2(A_2)| \geq M + 2$, and A_2 is bounded. Thus we define inductively a sequence $(A_n)_{n \in \mathbb{N}}$ such that

$$A_n \subset F - \bigcup_{i=1}^{n-1} A_i$$

and a sequence $\mu_n \in \Lambda$ such that $|\mu_n(A_n)| \geq M + n$. Now

$$|\mu_n(A_n)| \leq \sum_{i \in \mathbb{N}} |\mu_n(A_i)| \leq \sup_{\mu \in \Lambda} \sum_{i \in \mathbb{N}} |\mu(A_i)| < + \infty$$

which proves that the supposition that there are unbounded sets is contradictory.

This completes the proof of Theorem 0.4 and Proposition 0.5.

Theorem 0.3 and 0.4 may be combined in an abstract result as follows.

0.7 PROPOSITION. _Let_ E _be a quasi-complete locally convex space with the following properties_:

a) _A subspace_ G _of_ E' _is_ $\sigma(E', E)$ _closed if its intersection with bounded closed subsets is closed._

b) _Any continuous linear mapping from_ ℓ^∞ _to_ E _is weakly compact._ _Then any total subset of_ E' _possesses the Orlicz property._

Frechet spaces have the property a) and if E is either separable or weakly Σ-complete E possesses property b). Indeed in the separable case this is a result due to Grothendieck [1] p. 168. When E is weakly Σ-complete any continuous linear map from a space C(K) to E is weakly compact, (for an arbitrary compact K, not only for a Stone space). This was proved by Pełczynski [1] p. 219 for Banach spaces E, and independently by Thomas [1] for arbitrary quasi-complete spaces. [1]

For reference we mention the following generalization of Orlicz Theorem, (cf., Thomas [1] (Appendix II)).

0.8 THEOREM. Let E be a quasicomplete locally convex space and let H be a total subset of E', such that E admits a fundamental system of zero-neighborhoods which are closed for the topology $\sigma(E, H)$, and such that E is essentially separable relative to the topology $\sigma(E, H)$. Then H possesses the Orlicz property.

In this article we shall use the results from this paragraph in the following way: Let E be a quasi-complete locally convex space, contained in a locally convex space F with continuous injection

$$E \hookrightarrow F .$$

Let $m : \mathcal{A} \to F$ be a countably additive set function defined on a σ-algebra $\mathcal{A}$. If it happens that $m(A) \in E$ for all $A \in \mathcal{A}$, and if the topology induced by F on E possesses the Orlicz property, the map $m : \mathcal{A} \to E$ is also countably additive. Recapitulating the previous results we can say that the topology induced by F on E has the Orlicz property in the

[1] In the article by Pełczynski the author describes a weakly Σ-complete Banach space as a space not containing a subspace isomorphic to c_0, by virtue of a remarkable result of Bessaga and Pełczynski [1] p. 160, according to which a Banach space is weakly Σ-complete if (and only if) there is no subspace isomorphic to c_0. However we do not know if this equivalence holds for arbitrary quasi-complete spaces, and Pełczynski only uses the fact that the space is weakly Σ-complete.

following cases:

1) E is metrizable and weakly Σ-complete (in particular metrizable and weakly sequentially complete or reflexive)

2) E is separable (or essentially separable with respect to the topology induced by F) and furthermore at least one of the following conditions is satisfied:

a) E is metrizable, b) E admits a fundamental system of neighborhoods which are closed for the topology induced by F.

§1. Summable functions. The concept of summable family may be generalized as follows: let T be a topological Hausdorff space and let ν be a positive Radon measure on T.[1] A function $f : T \to X$ to a topological space X is said to be ν-measurable when for every compact K, and every $\varepsilon > 0$, there is a compact set $K' \subset K$, with $\nu(K - K') \leq \varepsilon$, such that the restriction $f|_{K'}$ of f to K' is continuous.

A scalar valued function f is ν-integrable when it is ν-measurable and when

$$\sup_K \int_K |f| \, d\nu < +\infty ,$$

the supremum being equal to $\int |f| \, d\nu$. In this case, when C_f is the set compact subsets $K \subset T$ such that $f|_K$ is continuous we also have

$$\int |f| \, d\nu = \sup_{K \in C_f} \int_K |f| \, d\nu$$

and for $\int f \, d\nu$, defined by decomposing f into non-negative components, we have

$$\int f \, d\nu = \lim_{K \in C_f} \int_K f \, d\nu$$

the limit being taken with respect to the filtering increasing family of compact sets $K \in C_f$. Conversely if this limit exists it follows that the scalar function f is ν-integrable.

---

[1] That is, a positive measure on the Borel σ algebra of T (smallest σ-algebra containing the open sets), such that every point has a neighborhood of finite measure and such that $\nu(A) = \sup_{K \subset A} \nu(K)$, for K compact, A Borel.

More generally let $\vec{f} : T \to E$ be a ν-measurable function with values in a quasi-complete locally convex space E. Let C_f be the set of compact subsets of T such that $\vec{f}|_K$ is continuous. For $K \in C_f$

$$\int_K \vec{f}\, d\nu$$

is an element of E which can be defined by Riemann sums (i. e. , by uniform approximation by simple functions) and which is characterized by the relations

$$\left\langle \int_K \vec{f}\, d\nu\, ,\, x' \right\rangle = \int_K \left\langle \vec{f}(t)\, ,\, x' \right\rangle d\nu(t)$$

for all linear forms $x' \in E'$.[1]

1. 1 PROPOSITION. Let $\vec{f} : T \to E$ be a ν-measurable function. The following properties are equivalent:

a)

$$\lim_{K \in C_f} \int_K \vec{f}\, d\nu \qquad \text{exists in } E.$$

b) (Cauchy criterium.) For any zero neighborhood V, there exists a compact set $K_0 \in C_f$ such that for $K \in C_f$, $K \cap K_0 = \emptyset$,

$$\int_K \vec{f}\, d\nu \in V \; .$$

c) For any sequence $(K_n)_{n \in N}$ of disjoint compact sets belonging to C_f, the sequence

1 See Bourbaki [2] Chapter 6, §1, Proposition 8.

$$\left(\int_{K_n} \vec{f} \, d\nu \right)_{n \in N}$$

<u>is summable in</u> E.

d) (Orlicz-Pettis condition.)

i) $\int |\langle \vec{f}(t), x' \rangle| \, d\nu(t) < +\infty$ <u>for all</u> $x' \in E'$.

ii) <u>For any Borel set</u> A <u>there exists</u>

$$\int_A \in E$$

<u>such that</u>

$$\left\langle \int_A, x' \right\rangle = \int_A \langle \vec{f}(t), x' \rangle \, d\nu(t)$$

<u>for all</u> $x' \in E'$.[1]

<u>Proof</u>. In this proof all sets denoted by K, K_0, K_i etc. will be compact sets belonging to C_f. a) implies d): let

$$S = \lim_K \int_K \vec{f} \, d\nu$$

and let $V = B^0$ be a zero-neighborhood in E, polar of a bounded set $B \subset E'$, and suppose

[1] In d) instead of supposing that $\int_A$ exists for all Borel sets it would be sufficient to suppose that $\int_\omega$ exists for all open sets ω. Indeed the map $\chi_\omega \to \int_\omega$ could be extended by continuity to the set of bounded functions which are uniform limits of characteristic functions of open sets, furnishing a bounded extendible Radon measure (cf., Thomas [1] §2, Theorem 2.2 bis.)

$$\int_K \vec{f}\, d\nu - S \in \frac{1}{2} V \qquad \text{for} \quad K \supset K_0 .$$

Then

$$\int_K \vec{f}\, d\nu - \int_{K_0} \vec{f}\, d\nu \in V \qquad \text{for} \quad K \supset K_0 ,$$

hence

$$\left| \int_{K - K_0} \langle \vec{f}, x' \rangle \, d\nu \right| \le 1 \qquad \text{for all} \quad x' \in B .$$

This implies first that

$$\int_{\complement K_0} | \langle \vec{f}, x' \rangle | \, d\nu \le 2$$

whence $\langle \vec{f}, x' \rangle$ is ν-integrable, and second that for any ν-measurable set A,

$$\left| \int_{A \cap (K - K_0)} \langle \vec{f}, x' \rangle \, d\nu \right| \le 1 \qquad \text{for all} \quad x' \in B ,$$

whence

$$\int_{A \cap K} \vec{f}\, d\nu - \int_{A \cap K_0} \vec{f}\, d\nu \in V .$$

Therefore it follows from the fact that E is quasi-complete that the limit

$$\lim_K \int_{A \cap K} \vec{f}\, d\nu$$

exists.[1] If we put this equal to $\int_A$ we have

$$\left\langle \int_A , x' \right\rangle = \int_A \langle \vec{f}, x' \rangle \, d\nu ,$$

in particular when A is a Borel set, whence d).

d) implies c). Let $(K_n)_{n\in N}$ be a sequence of disjoint compact sets belonging to C_f, and put

$$A_P = \bigcup_{n\in P} K_n \qquad \text{for} \quad P \subset N .$$

Then

$$\left\langle \int_{A_P} , x' \right\rangle = \sum_{n\in P} \int_{K_n} \langle \vec{f}, x' \rangle \, d\nu = \sum_{n\in P} \left\langle \int_{K_n} \vec{f} \, d\nu , x' \right\rangle$$

and Orłicz theorem implies that the sequence

$$\left(\int_{K_n} \vec{f} \, d\nu \right)_{n\in N}$$

is summable.

c) implies b). For if we supposed b) not to be valid there would exist a zero-neighborhood V and a sequence of disjoint compact sets $K_n \in C_f$ such that $\int_{K_n} \vec{f} \, d\nu \notin V$, contradicting c).

Finally b) implies a). Let $V = B^0$ be a zero-neighborhood and suppose $\int_K \vec{f} \, d\nu \in V$ for all $K \cap K_0 = \emptyset$, i.e.,

---

1 A priori the limits $\lim_K \int_{K\cap A} \vec{f} \, d\nu$ exist only in the completion $\hat{E}$ of E. However we have $|\langle \int_{A\cap K} \vec{f} \, d\nu, x' \rangle| \leq \int_A |\langle f, x' \rangle| \, d\nu$ which shows that the integrals $\int_{A\cap K} \vec{f} \, d\nu$ form a bounded family hence converge in E.

$$\left| \int_K \langle \vec{f}, x' \rangle \, d\nu \right| \leq 1 \qquad \text{for all} \quad x' \in B \ .$$

It follows that

$$\left| \int_{K_1 - K_0} \langle \vec{f}, x' \rangle \, d\nu \right| \leq 1 \qquad \text{for all} \quad x' \in B \ \text{ and all } \ K_1 \supset K_0,$$

that is

$$\int_{K_1} \vec{f} \, d\nu - \int_{K_0} \vec{f} \, d\nu \in V \qquad \text{for all} \quad K_1 \supset K_0 \ .$$

The existence of the limit follows from the fact that E is quasi-complete. (See footnote on p. 28.)

1.2 DEFINITION. A ν-measurable function for which the above conditions are satisfied is said to be ν-summable. The elements $\int_A$ are denoted by $\int_A \vec{f} \, d\nu$. Thus

$$\int_A \vec{f} \, d\nu = \lim_{K \in C_f} \int_{A \cap K} \vec{f} \, d\nu$$

and

$$\left\langle \int_A \vec{f} \, d\nu , x' \right\rangle = \int_A \langle \vec{f}, x' \rangle \, d\nu$$

It follows from condition d) that the sum of two ν-summable functions is still ν-summable and that

$$\int \vec{f} + \vec{g} \, d\nu = \int \vec{f} \, d\nu + \int \vec{g} \, d\nu \ .$$

Since $\lim_K \int_{A \cap K} \vec{f}\, d\nu$ exists it follows follows from condition a) that $\chi_A f$ is again ν-summable. More generally, let g be any ν-measurable bounded scalar function. Then $g\vec{f}$ is ν-summable. This is clear when g is a simple function and since

$$\left| \left\langle \int g\vec{f}\, d\nu , x' \right\rangle \right| \leq \| g \|_\infty \int |\langle \vec{f}, x' \rangle| \; d\nu \;,$$

the mapping $g \to \int g\vec{f}\, d\nu$ can be extended by continuity to the set of all bounded ν-measurable functions, and the extension is characterized by the relation

$$\left\langle \int g\vec{f}\, d\nu , x' \right\rangle = \int \langle g\vec{f}, x' \rangle \, d\nu \;,$$

and replacing g by $\chi_A g$ we have:

$$\left\langle \int_A g\vec{f}\, d\nu , x' \right\rangle = \int_A \langle g\vec{f}, x' \rangle \, d\nu$$

for all Borel sets A, so that $g\vec{f}$ is ν-summable by condition d) and the integral of $g\vec{f}$ coincides with the extension by continuity.

When E is a Banach space a function $\vec{f}$ is said to be ν-integrable or absolutely ν-summable when $\vec{f}$ is ν-measurable and $\int |\vec{f}|\, d\nu < +\infty$. Such a function is a fortiori ν-summable. For instance condition c) is fulfilled, the sequence being even absolutely summable:

$$\sum_n \left| \int_{K_n} \vec{f}\, d\nu \right| \leq \sum_n \int_{K_n} |\vec{f}|\, d\nu \leq \int |\vec{f}|\, d\nu$$

Similarly if E is a quasi-complete locally convex space any ν-measurable function $\vec{f}$ for which $\int |\vec{f}(t)|_p \, d\nu(t) < +\infty$ for all continuous semi-norms $p : x \to |x|_p$, is ν-summable, and called absolutely ν-summable.

1.3 PROPOSITION. Let E be a quasi-complete locally convex space which is weakly Σ-complete.[1] Then a function $\vec{f} : T \to E$ is ν-summable if and only if it is ν-measurable and $\int |\langle \vec{f}, x' \rangle| \, d\nu < +\infty$ for all $x' \in E'$.

Proof. Let $(K_n)_{n \in N}$ be a sequence of disjoint compact sets belonging to C_f. Then

$$\sum_n \left| \left\langle \int_{K_n} \vec{f} \, d\nu , x' \right\rangle \right| \leq \int |\langle \vec{f}, x' \rangle| \, d\nu < \infty$$

for all x'. Hence

$$\left(\int_{K_n} \vec{f} \, d\nu \right)_{n \in N}$$

is a summable sequence by Orlicz theorem and $\vec{f}$ is summable by condition c) Proposition 1.1.

[1] Cf/. Definition in Introduction. (Page 5).

§2. Radon measures with densities. In this paragraph and the following we shall suppose T to be a compact Hausdorff space.[1] Let $\vec{\mu} : C(T) \to E$ be a Radon mapping, and let ν be a positive Radon measure defined on T.

2.1 DEFINITION. A function $\vec{f} : T \to E$ is said to be a density of $\vec{\mu}$ with respect to ν when:

a) $\vec{f}$ is ν-measurable.

b) $\langle \vec{f}, x' \rangle$ is ν-integrable for all $x' \in E'$.

c) $\langle \vec{\mu}(\varphi), x' \rangle = \int \langle \vec{f}, x' \rangle \varphi \, d\nu$ for all $x' \in E'$ and all $\varphi \in C(T)$.

In this case we write $\vec{\mu}(\varphi) = \int \varphi \vec{f} \, d\nu$ or briefly $\vec{\mu} = \vec{f} \cdot \nu$.

2.2 PROPOSTION. Let $\vec{\mu} = \vec{f} \cdot \nu$. Then $\vec{\mu}$ is a Radon measure if and only if $\vec{f}$ is ν-summable. When this is the case a scalar function g is $\vec{\mu}$-integrable if and only if $g\vec{f}$ is ν-summable, and then

$$\int g \, d\vec{\mu} = \int g\vec{f} \, d\nu .$$

If furthermore E is a Banach space $\vec{\mu}$ is of bounded variation if and only if $\vec{f}$ is absolutely summable, and then the variation of $\vec{\mu}$ is given by $|\vec{\mu}|(\varphi) = \int \varphi |\vec{f}| \, d\nu$.

Proof. The first assertion is clear from the definition of Radon measures (cf., Introduction) and condition d) in Proposition 1.1.

---

[1] For many examples in analysis the basic space T, an open subset in Euclidean space, is locally compact. We leave to the reader to make the appropriate adjustments, replacing summable by locally summable etc.

The second assertion well known when both $\vec{\mu}$ and ν are scalar valued.[1] Let $\vec{\mu}_{x'}$ be the measure obtained by composition of $\vec{\mu}$ with the linear form x'. Then $\vec{\mu}_{x'} = \langle f, x' \rangle \nu$. It follows that $g\vec{f}$ is ν-summable if and only if g is $\vec{\mu}_{x'}$-integrable for all $x' \in E'$ and for every Borel set A there exists an element $\vec{\mu}(\chi_A g) \in E$ such that $\langle \vec{\mu}(\chi_A g), x' \rangle = \int_A g \, d\vec{\mu}_{x'}$ for all $x' \in E'$. But this is equivalent to the fact that g is $\vec{\mu}$-integrable, with

$$\int_A g \, d\vec{\mu} = \vec{\mu}(\chi_A g) \quad ,$$

by Theorem 3.11 or 3.20 in Thomas [1].[2]

The last assertion is well known and has been proved by Bourbaki [2], Chapter 6, Proposition 8, p. 40. A generalization of this will be given in paragraph 6.

2.3 LEMMA. Let T be a topological space and let ν be a positive Radon measure on T. Let $\vec{f} : T \to E$ be a ν-measurable function with values in a locally convex Hausdorff space E, and let B be a closed convex set in E. Suppose that for every closed half space

$$H_+ = \{x : \mathrm{Re} \langle x, x' \rangle \leq \alpha\}$$

containing B one has $\vec{f}(t) \in H_+$ ν-almost everywhere. Then $\vec{f}(t) \in B$ ν-almost everywhere.

---

1 Cf., Bourbaki [2] Chapter 5 (2nd Edition) §3. Theorem 1.

2 This result will not be used in the rest of this paper.

Proof. By the regularity of ν it suffices to prove this when T is compact which we assume. When $\vec{f}$ is continuous the hypothesis implies that $\vec{f}(t) \in H_+$ for all t belonging to the support of ν, and for all H_+; since by the Hahn-Banach theorem B is equal to the intersection of all half spaces H_+, this implies $\vec{f}(t) \in B$ for all t belonging to the support of ν, hence ν-almost everywhere.

In the general case T can be written as a disjoint countable union $T = N + \Sigma_n K_n$ where the restriction of $\vec{f}$ to the compact sets K_n are continuous and where N is ν-negligible. It follows that $\vec{f}(t) \in B$ for almost all $t \in K_n$, hence for almost all $t \in T$.

In particular if $x' \circ \vec{f}(t) = 0$ a.e. for all x', it follows that $\vec{f}(t) = 0$ a.e.

2.4 INCLUSION LEMMA. Let E be a quasi-complete locally convex space and let $\vec{\mu} : C(T) \to E$ be a Radon mapping having a density $\vec{f}$ with respect to ν.

a) Let B be a closed absolutely convex subset of E.[1] Then $\vec{\mu}(\varphi) = \int \varphi \vec{f} \, d\nu \in B$ for all $\varphi \in C(T)$ with $\int |\varphi| \, d\nu \leq 1$, if and only if $\vec{f}(t) \in B$ ν-almost everywhere.

b) Let P be a closed convex cone. Then $\int \varphi \vec{f} \, d\nu \in P$ for all continuous $\varphi \geq 0$, if and only if $\vec{f}(t) \in P$ ν-a.e.

c) Let B be a closed convex set and suppose $\vec{f}$ to be ν-summable. Then

---

[1] That is, is convex and such that $\lambda x \in B$ for $x \in B$ and $|\lambda| \leq 1$.

$$\frac{1}{\nu(A)} \int_A \vec{f}\, d\nu \in B$$

<u>for all Borel sets</u> A <u>with</u> $\nu(A) > 0$ <u>if and only if</u> $\vec{f}(t) \in B$ ν-a.e.

d) <u>Let</u> P <u>be a closed convex cone and suppose</u> $\vec{f}$ <u>to be</u> ν-<u>summable</u>. <u>Then</u> $\int_A \vec{f}\, d\nu \in P$ <u>for all Borel sets</u> A <u>if and only if</u> $\vec{f}(t) \in P$ ν-a.e.

<u>Proof.</u> By the previous lemma the proof is immediately reduced to the case where B or P are closed half spaces, in which case the result is obvious. For instance for a),

$$\left| \left\langle \int \vec{f} \varphi\, d\nu , x' \right\rangle \right| \leq 1 \qquad \text{for all } \varphi \in C(T)$$

with

$$\int |\varphi|\, d\nu \leq 1$$

is equivalent to $|\langle \vec{f}(t), x' \rangle| \leq 1$ for almost all t.

2.5 COROLLARY. <u>If</u> $\vec{\mu}$ <u>is a Radon mapping (resp. measure) with density</u> $\vec{f}$ <u>with respect to</u> ν, <u>and if</u> E_0 <u>is a closed subspace of</u> E, $\vec{\mu}(\varphi) \in E_0$ <u>for all</u> $\varphi \in C(T)$ (<u>resp.</u> $\int_A \vec{f}\, d\nu \in E_0$ <u>for all Borel sets</u> A) <u>if and only if</u> $\vec{f}(t) \in E_0$ ν - a.e.

2.6 COROLLARY. (<u>Uniqueness of densities). If</u> $\vec{\mu} = \vec{f}_1 \cdot \nu = \vec{f}_2 \cdot \nu$ $\vec{f}_1(t) = \vec{f}_2(t)$ ν-<u>almost everywhere. Conversely if</u> $\vec{\mu} = \vec{f}_1 \cdot \nu$ <u>and</u> $\vec{f}_2(t) = \vec{f}_1(t)$ ν-a.e. $\vec{f}_2$ <u>is also a density of</u> $\vec{\mu}$ <u>with respect to</u> ν.

Thus a density can properly be regarded as a class of ν-measurable functions equal ν-almost everywhere.

Proof. If

$$\int \vec{f}_1 - \vec{f}_2 \varphi \, d\nu = 0 \qquad \text{for all} \qquad \varphi \in C(T) \vec{f}_1(t) = \vec{f}_2(t)$$

ν-a.e. by the previous corollary with $E_0 = \{0\}$. The converse results immediately from the definition.

2.7 COROLLARY. Let $\vec{\mu}$ be a Radon mapping and suppose

$$\vec{\mu} = \vec{f}_1 \cdot \nu_1 = \vec{f}_2 \cdot \nu_2 \; .$$

Then if $S_i = \{t : \vec{f}_i(t) \neq 0\}$ the measures $\chi_{S_1} \cdot \nu_1$ and $\chi_{S_2} \cdot \nu_2$ are equivalent (i.e., have same sets of measure zero) and if

$$\chi_{S_2} \cdot \nu_2 = \rho_{1,2} \chi_{S_1} \cdot \nu_1$$

we have

$$\vec{f}_1(t) = \rho_{1,2}(t) \vec{f}_2(t) \quad \nu_1 \text{ a.e.}$$

In particular if $\vec{f}_2(t)$ belongs ν_2-almost everywhere to a (not necessarily closed) linear subspace, $\vec{f}_1(t)$ belongs ν_1-almost everywhere to the same subspace.

2.8 PROPOSITION. Let $\vec{\mu} : C(T) \to E$ be a Radon map and suppose $\vec{\mu} = \vec{f} \cdot \nu$. Let $u : E \to F$ be a continuous linear map. Then $u \circ \vec{\mu} = u \circ \vec{f} \cdot \nu$.

In particular if E is contained in F with continuous injection $j : E \hookrightarrow F$, $j \circ \vec{\mu} = j \circ \vec{f} \cdot \nu$. By a simplification of the language we shall describe this saying that if $\vec{\mu}$ has an E-valued density $\vec{f}$ this density is a fortiori an F-valued density. It follows from the uniqueness property (2.6) that if $\vec{\mu}$ has an F-valued density $\vec{f}$, $\vec{\mu}$ can only have an E-valued density $\vec{g}$ if $\vec{f}(t) = \vec{g}(t)$ (strictly $\vec{f}(t) = j \circ \vec{g}(t)$) ν-almost everywhere, which implies that $\vec{f}(t) \in E$ ν-almost everywhere. This justifies the remarks in the introduction.

<u>Absolute continuity and weak absolute continuity.</u> Let $\vec{\mu}: C(T) \to E$ be a Radon measure. Then the following conditions are equivalent:

a) $$\lim_{\nu(A) \to 0} \vec{\mu}(A) = 0 \ ;$$

b) $$\nu(A) = 0 \quad \text{implies} \quad \vec{\mu}(A) = 0 \ .$$

When they are satisfied $\vec{\mu}$ is said to be <u>absolutely continuous with respect to</u> ν, or briefly ν-<u>continuous</u>.

These conditions are also equivalent to the following:

c) $x' \circ \vec{\mu}$ is ν-continuous for all $x' \in E'$.

The latter condition also makes sense for an arbitrary Radon mapping $\vec{\mu}$, and a Radon mapping satisfying c) will be said to be <u>weakly</u> ν-<u>continuous</u>. Thus a Radon mapping is weakly ν-continuous if the associated Radon measure, with values in the weak completion E'^* of E, is ν-continuous. A weakly ν-continuous Radon mapping is ν-continuous if and only if it is a Radon measure. Thus for Radon measures absolute continuity and weak absolute continuity are the same.

It has been shown by Bartle, Dunford and Schwartz [1], that any Radon measure $\vec{\mu} : C(T) \to E$ with values in a metrizable space is absolutely continuous with respect to some positive Radon measure ν. When E is a Banach space this results from the fact that the set of scalar measures $\{x' \circ \vec{\mu}\}_{|x'| \leq 1}$ is weakly relatively compact in $M(T) = C(T)'$.

Evidently a Radon mapping with density $\vec{f}$ with respect to ν, is weakly ν-continuous. That the converse is false, even when $\vec{\mu}$ is a Radon measure, i. e., that absolute continuity does not imply the existence of a density, will be clear in the light of the following theorem:

2.9 THEOREM. _Let_ E _and_ F _be quasi-complete locally convex spaces such that_ E _is included in_ F, _with continuous injection_ $E \subsetneq F$. _Suppose furthermore that at least one of the following hypotheses is satisfied:_

a) E _is metrizable and separable (or essentially separable with respect to_ F).[1]

b) E _is metrizable and weakly sequentially complete (or weakly_ Σ_-complete)._

c) E _has a fundamental system of zero neighborhoods which are closed for the topology induced by_ F, _and_ E _is essentially separable with respect to_ F.[1]

Let $\vec{f} : T \to F$ _be a_ ν_-summable function such that for every Borel set_ A _we have_

$$\int_A \vec{f} \, d\nu \in E .$$

---

[1] That is, every countable subset of E is contained in a separable subspace of E which is closed for the topology induced by F.

Then the E*-valued function* $\vec{\mu}$ *defined by*

$$\vec{\mu}(A) = \int_A \vec{f}\, d\nu$$

is a ν*-continuous Radon measure. Furthermore, a function* g *is* $\vec{\mu}$*-integrable if and only if* $g\vec{f}$ *is* ν*-summable and* $\int_A g\vec{f}\, d\nu \in E$ *for all Borel sets* A .

Proof of the first assertion. It follows from the results in §0 that the topology induced by F on E possesses the Orlicz property. Hence $\vec{\mu}$ is a countably additive set function. Since $\nu(A) = 0$ implies $\vec{\mu}(A) = 0$, $\vec{\mu}$ is ν-continuous, in particular, since ν is regular $\vec{\mu}$ is regular, hence a Radon measure. We shall not give the proof of the second assertion (which will not be used hereafter) except under hypothesis c). It then follows from Proposition 2.2 (applied with F instead of E) and Theorem 3.20 in Thomas [1].

It will be seen from the examples that under the hypotheses of the above theorem it can very well happen that $\vec{f}(t)$ does not belong to E for almost every t, so that $\vec{\mu}$ does not have a density with respect to ν, nor (by 2.7) with respect to any other positive measure. However, Theorem 2.9 shows that in a certain sense the function $\vec{f}$ is a good substitute for an E-valued density. For example one has the following dominated convergence property: if g_n is ν-measurable and $g_n(t)$ tends to $g(t)$ ν-almost everywhere, with $|g_n| \le h$ where $h\vec{f}$ is ν-summable and $\int_A h\vec{f}\, d\nu \in E$ for all Borel sets A , $\int g_n \vec{f}\, d\nu$ tends to $\int g\vec{f}\, d\nu$ in E (and not just for the topology induced by F).

The situation described in Theorem 2.9 is a particular case of the following general situation which we shall study in the remainder of this article: E and F are quasi-complete locally convex spaces with a given continuous imbedding $j : E \hookrightarrow F$, by which we identify E to a linear subspace of F. $\vec{\mu} : C(T) \to E$ is a Radon mapping such that $j \circ \vec{\mu}$ has a density $\vec{f}$ with respect to ν. We shall describe this situation saying that $\vec{\mu}$ has a density in F. $E \hookrightarrow F$, $\vec{\mu}(\varphi) = \int \vec{f} \varphi \, d\nu$ in F.

When $\vec{\mu} : C(T) \to E$ is a Radon measure it follows that $\vec{\mu}$ is ν-continuous, and in general we shall still assume that $\vec{\mu}$ is weakly ν-continuous.

Following is a list of examples illustrating the above considerations:

EXAMPLES

2.10a. Let $E = c_0$ be the Banach space of sequences tending to zero, and let $F = \mathbb{C}^{\mathbb{N}}$ be the space of all sequences. Let ν be a positive Radon measure on a compact space T, and let $(f_n)_{n \in N}$ be a sequence of scalar ν-integrable functions such that

$$\lim_{n \to \infty} \int \varphi f_n \, d\nu = 0 \qquad \text{for all} \quad \varphi \in C(T) .$$

Put $\vec{\mu}(\varphi) = \{ \int \varphi f_n \, d\nu \}_{n \in N}$. Then $\vec{\mu} : C(T) \to c_0$ is a Radon mapping (closed graph theorem) and $\vec{\mu}(\varphi) = \int \varphi \vec{f} \, d\nu$ in $\mathbb{C}^{\mathbb{N}}$, where $\vec{f}(t) = (f_n(t))_{n \in N}$. However the hypothesis does not imply $\vec{f}(t) \in c_0$ for almost all t.

Specifically let for instance $T = [0, 1]$ and let ν be Lebesgue measure. Then with $f_n(t) = \overline{e}^{2\pi int}$ we have $|f_n(t)| = 1$ for all t

and n. In this case $\vec{\mu}$ is a Radon measure

$$\lim_{n \to \infty} \int_A f_n \, dt = 0$$

and even of bounded variation:

$$|\vec{\mu}(\varphi)|_{c_0} \leq \int |\varphi| \, dt .$$

2.10b. Consider now on the contrary a sequence $(f_n)_{n \in N}$ of ν-integrable functions such that

a) $$\int |f_n| \, d\nu = 1$$

b) $$\lim_{n \to \infty} f_n(t) = 0$$

c) $$\lim_{n \to \infty} \int \varphi f_n \, d\nu = 0 \qquad \text{for all} \quad \varphi \in C(T)$$

(e.g., with $T = [-1, +1]$, f_n the continuous odd function linear in

$$\left[0, \frac{1}{2n}\right], \quad \left[\frac{1}{2n}, \frac{1}{n}\right], \quad \left[\frac{1}{n}, 1\right]$$

such that

$$f_n(0) = f_n\left(\frac{1}{n}\right) = f_n(1) = 0 ,$$

and $f_n\left(\frac{1}{2n}\right) = n$). Let

$$\vec{\mu}(\varphi) = \left(\int \varphi f_n \, d\nu\right)_{n \in N} \in c_0 .$$

Then $\vec{f}(t) = (f_n(t))_{n\in N}$ is a density of $\vec{\mu}$ with respect to ν. Indeed for each $x' \in \ell^1 = c_0'$ we have

$$\langle \vec{f}(t), x' \rangle = \sum_n x_n' f_n(t)$$

so that $\langle \vec{f}, x' \rangle$ is integrable, and c_0 being separable, $\vec{f}$ is measurable (cf., Proposition 3.2), and furthermore

$$\langle \vec{\mu}(\varphi), x' \rangle = \sum_n \int \varphi f_n \, d\nu \, x_n' = \int \varphi \langle \vec{f}, x' \rangle \, d\nu .$$

However $\vec{\mu}$ is not a Radon measure, that is, $\vec{f}$ is not summable, for in the contrary case we would have

$$\lim_{n\to\infty} \int_A f_n \, d\nu = 0$$

for all Borel sets A that is, f_n would tend weakly to zero in $L^1(\nu)$, and since $f_n(t)$ tends to zero this would imply

$$\lim_{n\to\infty} \int |f_n| \, d\nu = 0$$

contrary to the hypothesis.

Thus a Radon mapping with a density is not necessarily a Radon measure.

2.11 Example. Let $T = \mathbb{R}$ the set of real numbers, and let ν be Lebesgue measure. Let $E = L^2(\nu)$ and $F = \mathscr{S}'(\mathbb{R})$ the space of temperate distributions. Let $\vec{\mu}(\varphi)$ be the Fourier transform of the continuous function with compact support, φ. Then $\vec{\mu}:\mathscr{K}(T)\to L^2$ is a Radon measure. Let $\vec{f}(t)$ be the function $\xi \to \bar{e}^{2\pi i\xi t}$. Then $\vec{f}: T\to \mathscr{S}'$ is locally summable and $\vec{\mu}(\varphi) = \int \varphi \vec{f}\, dt$ in $\mathscr{S}'$. However we do not have $\vec{f}(t) \in L^2$.

One can vary this example by taking for instance $E = C_0(\mathbb{R})$, in which case $\vec{\mu}$ is of locally bounded variation, or $F = L^1_{loc}(\mathbb{R})$ if one prefers to remain within the realm of functions.

2.12 Example. a) Let T be the circle group (or some other locally compact group) and let $\nu\ (= dt)$ be a Haar measure. Let $u \in L^1(T,\nu)$ and let $\vec{f}(t) = \tau_t u$ be the translate of u by t. Then $\vec{f}: T \to L^1$ is a continuous function. Let $\vec{\mu}(\varphi) = \int \vec{f}(t)\varphi(t)\, dt$. Then $\vec{\mu}(\varphi) = u * \varphi$ (convolution product) hence $\vec{\mu}(\varphi) \in C(T)$ for all $\varphi \in C(T)$ and $\vec{\mu}: C(T) \to C(T)$ is a Radon measure, with $\int g\, d\vec{\mu} = g * u$ for all bounded measurable functions g, $\vec{\mu}$ is ν-continuous and has a density in $L^1(T)$, but $\vec{f}(t) \in C(T)$ for almost all t if and only if $u \in C(T)$. Thus we have the situation described above with $E = C(T)$, $F = L^1(T)$.

2.12 b) More generally let $u \in M(T)$ be a measure and take F to be the space $M(T)$ equipped with the vague (weak*) topology $\sigma(M,C)$. Then $t \to \vec{f}(t) = \tau_t u$ is a summable function and

$$\vec{\mu}(\varphi) = \int \varphi \vec{f}\, dt \in L^1(T)$$

for all bounded measurable functions φ. Thus $\vec{\mu}: C(T) \to L^1(T)$ is a Radon measure with a density in F. Clearly $\vec{\mu}$ has a density in $E = L^1$ if and only if $u \in L^1(T)$.

When u is a measure we still have $\vec{\mu}(\varphi) = \varphi * u \in C(T)$ for all $\varphi \in C(T)$ and $\vec{\mu} : C(T) \to C(T)$ is a Radon mapping. In the above example we have seen that it is a Radon measure (i.e., a weakly compact mapping) when $u \in L^1(T)$. Conversely it is easy to see that $\vec{\mu} : C(T) \to C(T)$ is weakly compact only if $u \in L^1(T)$. We shall see later (Theorem 4.5) that this is a particular case of a general phenomenon. The particular case of example 12b where $u = \delta$ can be treated better in the following more general context:

2.13 Example. Let T be a compact space with a positive Radon measure ν. Let $F = M(T)$ be the space of measures equipped with the weak* topology $\sigma(M, C)$. Let $\vec{f}(t) = \delta_{(t)}$ be the Dirac measure at the point t. Then $\vec{f}$ is a continuous and hence ν-summable function. $\vec{\mu}(\varphi) = \int \varphi \vec{f}\, d\nu$ is of course the measure $\varphi \cdot \nu$ with density φ with respect to ν. Thus identifying $L^1(\nu)$ with the space of ν-continuous measures, $\vec{\mu} : C(T) \to L^1(\nu) \subsetneq M$ is a Radon measure with density in M.

If ν is dense so that $C(T) \subset L^1(\nu)$, we can also interpret $\vec{\mu}$ as a Radon map (here the identity) from $C(T)$ to $C(T)$, with density in $M(T)$.

2.14 Example. a) Greens function. Let $T = [a, b]$ and let D be a formally selfadjoint second order non singular differential operator with C^∞ coefficients, and let $Du = \varphi$, $u \in \mathscr{L}$ be some associated Sturm-Louisville problem with appropriate boundary conditions $\mathscr{L}$ ($u \in \mathscr{L}$, $Du = 0$ implies $u = 0$). Let $u = \vec{\mu}(\varphi)$ be the solution for $\varphi \in C[a, b]$. Then $\vec{\mu} : C[a, b] \to C^2[a, b]$ is a Radon mapping, and $\vec{\mu} : C[a, b] \to C^1[a, b]$ is a Radon measure. The existence of the Greens function (or Corollary 4.2) shows that $\vec{\mu}$ has a density $t \to G_t$ with values in $C[a, b]$

$$u = \int_a^b G_t \varphi(t)\, dt$$

or

$$u(x) = \int_a^b G(x,t)\varphi(t)\, dt \quad .$$

Since

$$\varphi = Du = \int_a^b DG_t \varphi(t)\, dt$$

we must have $DG_t = \delta_{(t)}$ a.e. which proves that by its very nature G_t cannot belong to $C^1[a,b]$.

2.14 b) Let Ω be an open subset of $\mathbb{R}^m$, $m \geq 3$. Let $\vec{\mu}(\varphi)$ be the solution of the problem $-\Delta u = \varphi\; u \in \mathscr{H}_0^1(\Omega)$. Then $\vec{\mu} : \mathscr{K}(\Omega) \to \mathscr{H}_0^1(\Omega)$ is a Radon measure and $\vec{\mu}(\varphi) = \int G_t \varphi(t)\, dt$ in the space $D'(\Omega)$. Since $-\Delta \vec{\mu}(\varphi) = \int -\Delta G_t \varphi(t)\, dt = \int \delta_{(t)} \varphi(t)\, dt$ the uniqueness property implies $-\Delta G_t = \delta_{(t)}$ a.e. (in fact everywhere) so that G_t cannot take its values in $\mathscr{H}_0^1$.

2.15 Example. Let Ω be a bounded open set in $\mathbb{R}^m$ and let $T = \partial\Omega$ be the compact boundary of Ω. Let $F = \mathrm{Harm}(\Omega)$ be the space of harmonic functions in Ω and let $\vec{\mu}(\varphi)$ be the solution of the Dirichlet problem (in generalized sense) $\Delta u = 0$, $u/\partial\Omega = \varphi$. Then $\vec{\mu} : C(T) \to \mathrm{Harm}(\Omega)$ is a Radon measure. If the boundary $T = \partial\Omega$ is sufficiently regular, with measure of surface area ν, we have

$$\vec{\mu}(\varphi) = \int P_t \varphi(t)\, d\nu(t)$$

where $t \to P_t \in \mathrm{Harm}(\Omega)$ is ν-summable, (the function $(t,s) \to P_t(s)$ on $T \times \Omega$ is called the Poisson kernel). Let $E = C(\overline{\Omega}) \cap \mathrm{Harm}(\Omega)$ with the obvious Banach space structure, then $\vec{\mu}: C(T) \to E$ is a Radon mapping but we do not have $P_t \in E$. Indeed if $R: E \to C(T)$ is the restriction mapping we would have $\varphi = R\vec{\mu}(\varphi) = \int RP_t \varphi(t)\, dt$ and consequently $RP_t = \delta_{(t)}$ a.e. which is absurd.

2.16 Example. Let T be a locally compact space with positive Radon measure ν. Let $u: L^P(\nu) \to L^P(\nu)$ $(1 \le p < +\infty)$ be a continuous linear operator, and let $\vec{\mu}: \mathscr{K}(T) \to L^p(\nu)$ be the "restriction" of u to the space of continuous functions with compact support. Then $\vec{\mu}$ is a ν-continuous Radon measure which completely determines u, since $\mathscr{K}$ is dense in L^p.[1] Suppose now that L^p is imbedded appropriately in some larger space, specifically suppose $T = \Omega$ is some open subset of $\mathbb{R}^m$, and consider the inclusion $L^p(\Omega) \subsetneq \mathscr{D}'(\Omega)$ where $\mathscr{D}'(\Omega)$ is the space of distributions in Ω. Then $\vec{\mu}(\varphi) = \int \varphi(t)\vec{f}(t)\, dt$ where $\vec{f}$ is locally summable with respect to Lebesgue measure. But one does not in general have $\vec{f}(t) \in L^p$. For instance when u is the identity map $\vec{f}(t) = \delta_{(t)}$, when $\Omega = \mathbb{R}^m$, $p = 2$, and u is Fourier transform, $\vec{f}(t) = e^{-2\pi i t.}$, when $\Omega = \mathbb{R}^m$ and u commutes with translations

$$\vec{\mu}(\varphi) = \varphi * T = \int \tau_t T \varphi(t)\, dt ,$$

where T is some distribution (a measure when $p = 1$, a pseudo-measure when $p = 2$). Whenever u is an invertible operator in L^p the

[1] More precisely $\mathscr{L}^p(\nu)$ is contained in $\mathscr{L}^1(\vec{\mu})$ and $u(\dot{f}) = \int f\, d\vec{\mu}$ where $\dot{f}$ is the class of f in L^p.

corresponding measure cannot have a density in L^p, for applying u^{-1} one would have $u^{-1}\vec{f}(t) = \delta_{(t)}$ which is absurd.

Thus it is seen that the phenomenon of a "density going out of the space" is in no wise pathological.

§ 3. Universally measurable mappings. In this section we display mostly known facts about measurable functions and densities for certain Radon measures for later reference. A proof of the Grothendieck-Phillips theorem is given based on the inclusion Lemma 2. 4, thus making the present article independent of previous proofs involving lifting.

Recall that given a metrizable topological space E , the limit of a pointwise almost everywhere convergent sequence of ν-measurable functions is ν-measurable.[1] When f is a ν-measurable function with values in a topological space E , the inverse image by f of a Borel set is a ν-measurable set. If furthermore E is metrizable and separable (or if f has essentially separable range) a function f with this property is ν-measurable. In particular when E is metrizable and separable any Borel function $f : T \to E$ is ν-measurable for an arbitrary positive Radon measure ν.[2] When f is a function $(f_n)_{n \in \mathbb{N}}$ with values in a countable product $\Pi_{n \in \mathbb{N}} E_n$, it is clear that f is ν-measurable if and only if the components f_n are ν-measurable. This enables one to reduce the search for densities with values in Fréchet spaces (which are closed subspaces of countable products of Banach spaces) to the case of Banach spaces by virtue of the following lemma.

[1] This is no longer the case when E is not metrizable. For example if $E = \mathbb{R}^I$ with $I = [0, 1]$. Let $f_n = (f_n^i)_{i \in I}$ with $f_n^i(t) = [1 - (t - i)^2]^n$ then f_n is continuous, takes its values in the compact space $[0, 1]^I$, and $\lim_n f_n(t) = f(t)$ with $f^i(t) = \delta_{i,t}$, so that f is not measurable with respect to Lebesgue measure.

[2] Cf. Bourbaki [2] Ch. IV, §5, Theorem 4.

3.1 LEMMA. Let E be a closed subspace of a countable product $\Pi_n E_n$ of quasi-complete locally convex spaces. Let $\vec{\mu} : C(T) \to E$ be a Radon mapping. Then $\vec{\mu}(\varphi) = (\vec{\mu}_n(\varphi))_{n \in \mathbb{N}}$ and $\vec{\mu}$ has a density with respect to ν if and only if each μ_n has a density with respect to ν.

Proof. Since $\vec{\mu}_n = \pi_n \circ \vec{\mu}$, with π_n the restriction to E of the projection map onto E_n, the necessity follows from Proposition 2.8, while the sufficiency follows from 2.5.

The following proposition generalizes a well known result according to which scalarly measurable functions with values in a separable Banach space are measurable.

3.2 PROPOSITION. Let E be a separable Banach space and let $(x'_n)_{n \in \mathbb{N}}$ be a sequence of continuous linear forms such that

$$|x| = \sup_n |\langle x, x'_n \rangle| \qquad \text{for all} \quad x \in E .$$

Let $f : T \to E$ be a function and let $f_n = x'_n \circ f$. Then f is ν-measurable if and only if f_n is ν-measurable for each n.

Proof. Let $j : E \to \mathbb{C}^{\mathbb{N}}$ be the mapping $j(x) = \{\langle x, x'_n \rangle\}_{n \in \mathbb{N}}$ and let $\mathcal{T}$ be the topology induced by j (or by the x'_n) on E. Since the map $t \to (f_n(t))_{n \in \mathbb{N}}$ is ν-measurable with values in $\mathbb{C}^{\mathbb{N}}$, it suffices to show that the identity $(E, \mathcal{T}) \to E$ is a Borel mapping. The unit ball of E is closed in E with respect to $\mathcal{T}$, hence every open set in E, countable union of closed balls is F_σ with respect to $\mathcal{T}$ and finally every Borel set in E is a Borel set in $\mathcal{T}$, which was to be proved.[1]

Remark. If one takes $E = c$, the space of convergent sequences, and for the x'_n the coordinate forms, the above proposition reduces to Bourbaki's version of Egoroff's theorem (cf., Bourbaki [2], Chapter IV, §5, Theorem 2, 2°).

The Theorem of Phillips-Grothendieck. This theorem essentially states the following: Let $\vec{\mu} : C(T) \to E$ be a Radon map with values in a Fréchet space E, and suppose there exists a positive Radon measure ν on T and a weakly compact set $B \subset E$ such that $\vec{\mu}(\varphi) \in B$ for all $\varphi \in C(T)$ with $\int |\varphi| \, d\nu \leq 1$. Then $\vec{\mu}$ has a density with respect to ν, (with values in B).

Proof. Since E is a closed subspace of a product of Banach spaces it is sufficient by Lemma 3.1 to consider the case where E is a Banach space. The hypothesis then states that $\vec{\mu}$ is not only a measure of bounded variation, but extends even to a weakly compact map $\vec{\mu} : L^1(\nu) \to E$. The fundamental theorem of Phillips ([1], 5.5) and Grothendieck ([1] p. 139) then implies that the image by $\vec{\mu}$ of the unit ball in $L^\infty(\nu)$ is relatively compact in E, hence separable, and it follows that the image of L^1 is a separable subspace of E. Thus we may and shall suppose that E is a separable Banach space. This crucial step having been taken the theorem now results directly from the inclusion Lemma 2.4 as follows. Let $(x'_n)_{n \in \mathbb{N}}$ be a sequence of continuous linear forms such that

---

[1] This proposition is a very particular case of a remarkable theorem due to P. A. Meyer and L. Schwartz to be quoted (cf. p. 51). The proof for an arbitrary separable Fréchet space E and any total sequence $(x'_n)_{n \in N}$ also follows from the theorem of Banach-Dieudonné, since the set of x' for which $x' \circ f$ is measurable is total and sequentially closed, hence equal to E'.

$$|x| = \sup_n |\langle x, x'_n \rangle| \qquad \text{for all} \quad x \in E,$$

and consider the injection $x \to \{\langle x, x'_n \rangle\}_{n \in \mathbb{N}} \in \mathbb{C}^{\mathbb{N}}$ by which we identify E with a linear subspace of $\mathbb{C}^{\mathbb{N}}$. We may assume that B is the closure in E of the set

$$\{\vec{\mu}(\varphi)\}_{\int|\varphi|\, d\nu \leq 1} \,.$$

Then B is absolutely convex and compact in the weak topology hence compact in $\mathbb{C}^{\mathbb{N}}$. It follows from the Lebesgue-Nikodym theorem for scalar measures that $\vec{\mu}$ has a density $\vec{f} = (f_n)$ with values in $\mathbb{C}^{\mathbb{N}}$, and be the inclusion Lemma 2.4 a, $\vec{f}(t) \in B$ for ν-almost all t, and after suitable modification, for all t. Then $\vec{f}$ is an E-valued measurable function by Proposition 3.2, and being bounded it is absolutely ν-summable. Since

$$\langle \vec{\mu}(\varphi), x'_n \rangle = \int f_n \varphi \, d\nu = \int \langle f, x'_n \rangle \varphi \, d\nu = \langle \int \vec{f} \varphi \, d\nu, x'_n \rangle$$

we have

$$\vec{\mu}(\varphi) = \int \varphi \vec{f} \, d\nu \quad ,$$

which terminates the proof.

The theorem of Dunford-Pettis in the present context can be proved in similar fashion (cf., 6.10).

Some universally measurable maps. Recall that a universally (Lusin) measurable map $f : X \to Y$ from a topological space X to a topological space Y, is a map such that for every compact set $K \subset X$ and any positive Radon measure ν on K, f is ν-measurable, i.e., for every $\epsilon > 0$, there is a compact set $K' \subset K$, with $\nu(K - K') \leq \epsilon$, such that the restriction of f to K' is continuous. The composition of two universally measurable maps is universally measurable, every continuous map is universally measurable; when Y is a separable metrizable space every Borel map $f : X \to Y$ is universally measurable. When X is a countable union of Borel or universally measurable subsets X_n, such that the restriction of f to each X_n is universally measurable, f is universally measurable. We shall often meet the situation where the sets X and Y coincide. Let X_1 and X_2 be two topological spaces with the same underlying set X. Then the following conditions are obviously equivalent:

a) The identity $X_2 \to X_1$ is universally measurable.

b) For every ν-measurable function $f : T \to X_2$ the function $f : T \to X_1$ is ν-measurable (T and ν arbitrary).

c) For every continuous function $f : T \to X_2$ the function $f : T \to X_1$ is universally measurable. (T arbitrary compact space).

The following is a corollary of the above mentioned theorem of Grothendieck and Phillips. Let E be a Fréchet space. Then the identity $\sigma(E, E') \to E$ is universally measurable. In other terms every E valued function $\vec{f} : T \to E$ which is weakly ν-measurable (i. e., ν-measurable with respect to the weak topology) is ν-measurable.

Indeed if a function $\vec{f} : K \to E$ is weakly continuous hence weakly bounded, and weakly ν-summable, it defines a Radon measure $\varphi \to \vec{\mu}(\varphi) = \int \varphi \vec{f} \, d\nu$ with values in the weak completion of E, such that for $\int |\varphi| \, d\nu \leq 1$ $\vec{\mu}(\varphi)$ belongs to the absolutely convex closed hull of $f(K)$, which is still weakly compact in E. Hence $\vec{\mu}$ is an E-valued Radon map which by the theorem of Grothendieck and Phillips has an E-valued density

$\vec{g}$ with respect to ν. Since $\vec{g}$ is a fortiori a density relatively to the weak topology it follows from the uniqueness that $\vec{f}(t) = \vec{g}(t)$ ν a. e., hence $\vec{f}$ is ν-measurable also.

This theorem is to be compared to the following remarkable theorem due to P. A. Meyer and L. Schwartz [1]: Let X_1 and X_2 be topological Hausdorff spaces with the same underlying set X. Suppose the identity $X_1 \to X_2$ is continuous and that X_1 is a K-analytic space (i. e., a continuous image of a $K_{\sigma\delta}$). Then the identity $X_2 \to X_1$ is universally measurable. When X_1 is a separable metrizable complete space (Polish space) X_1 is K-analytic. We shall also meet the particular case of this theorem when X_1 is K_σ, a countable union of compact sets, on each of which the topologies necessarily agree, so that the result is obvious in that case.

We shall apply the case of Polish spaces as follows: Let E be a <u>separable</u> Fréchet space and let $\mathscr{T}$ be an arbitrary Hausdorff topology which is weaker than the given topology. Then a function $\vec{f} : T \to E$ which is ν-measurable with respect to $\mathscr{T}$, is already ν-measurable with respect to the given topology.

The above mentioned case of the Meyer Schwartz theorem where X_1 is K_σ may be combined with the theorem of Phillips-Grothendieck to yield the following proposition:

3.3 PROPOSITION. *Let E be a reflexive Banach space, and let $\mathscr{T}$ be a locally convex Hausdorff topology which is weaker than the given topology. Then the identity $(E, \mathscr{T}) \to E$ is universally measurable.*

Proof. We may and shall replace $\mathscr{T}$ by the associated weak topology $\sigma(E, H)$ where H is the set of linear forms on E which are continuous with respect to $\mathscr{T}$. Since E is σ-compact for the topology $\sigma(E, E')$ the identity $\sigma(E, H) \to \sigma(E, E')$ is universally measurable, and the identity $\sigma(E, E') \to E$ being universally measurable by the theorem of Grothendieck-Phillips, the identity $\sigma(E, H) \to E$ is universally measurable by composition.

Remark. In the proof of Proposition 3. 2 it was shown that the identity $(E, \mathscr{T}) \to E$ was a Borel mapping hence universally measurable (E being a separable Banach space). The theorem of Meyer-Schwartz proves that for any total subset $H \subset E'$, the identity $\sigma(E, H) \to E$ is universally measurable. However it seems preferable to have a simple proof in the frequently occurring case described in Proposition 3. 2.

3. 4 Examples of measurable maps. Let $E = \ell^p(I)$ $1 \le p < +\infty$, or $E = C(K)$, K compact. Let $\mathscr{T}$ be the topology induced by $\mathbb{C}^I$ (resp. $\mathbb{C}^K$). Then the identity $(E, \mathscr{T}) \to E$ is universally measurable. When E is separable (i. e., I countable or K metrizable) the map is even a Borel map.

Proof. For C(K). A bounded set in C(K) which is compact for the topology of pointwise convergence is weakly compact (cf., Grothendieck [2], Chapter V, §3, Theorem 3). Thus if K is compact for the topology $\mathscr{T}$ of pointwise convergence K is σ-compact for the weak topology; hence the identity $(E, \mathscr{T}) \to \sigma(E, E')$ is universally measurable, and we conclude by the theorem of Phillips-Grothendieck. (Since every Banach space, E, is a subspace of a space C(K) with K unit ball of E', the weak topology being induced by the topology of pointwise convergence, this example can be considered as an equivalent rendering of the Grothendieck-Phillips measurability theorem). The case of ℓ^p with $1 < p < +\infty$ is a consequence of Proposition 3.3. In the case of ℓ^1 we proceed as follows. Put

$$p_J(x) = \sum_{i \in J} |x_i| .$$

Then

$$|x| = p_I(x) = \sup_{J \subset I} p_J(x)$$

where J is countable. Let $K \subset \ell^1(I)$ be compact for the pointwise topology and bounded in norm and let ν be a positive Radon measure on K. The functions p_J are lower semicontinuous on K, hence

$$\int |x| \, d\nu(x) = \sup_J \int p_J(x) \, d\nu(x) .$$

Consequently there exists a countable set J such that $\int |x| \, d\nu(x) = \int p_J(x) \, d\nu(x)$, whence $\int p_{\complement J}(x) \, d\nu(x) = 0$. Thus for ν-almost all $x \in K$ $x_i = 0$ for $i \in \complement J$, i.e., except for a set of measure zero K is situated in $\ell^1(J)$ where J is countable. Now when J is countable the identity $\ell^1_s(J) \to \ell^1(J)$ is a Borel map, hence measurable. Thus the identity $K \to \ell^1(I)$ is almost everywhere equal to a measurable map, hence measurable.

The case of ℓ^1 yields in particular the fact that the identity $\sigma(\ell^1, c_o) \to \ell^1$ is universally measurable, which is quite exceptional, normally if E' is the dual of a Banach space the identity $\sigma(E', E) \to E'$ is not universally measurable (e.g., $E = \ell^1$, or $C(K)$). However we note that the weak*topology is a kind of lower bound of topologies for which function may be measurable. Since E' is σ-compact for the topology $\sigma(E', E)$, it follows that any function $\vec{f} : T \to E'$ which is ν-measurable for some Hausdorff topology weaker than the weak*topology $\sigma(E', E)$ is in fact measurable for the weak*topology.

In the case of $\ell^\infty(I)$ we have the following:

3.5 EXAMPLE. Let $\ell^\infty_s(I)$ be the space $\ell^\infty(I)$ equipped with the topology of pointwise convergence, induced by $\mathbb{C}^I$, and let $\ell^\infty_\tau(I)$ be $\ell^\infty(I)$ endowed with the Mackey topology $\tau(\ell^\infty, \ell^1)$. Then the identity $\ell^\infty_s(I) \to \ell^\infty_\tau(I)$ is universally measurable.

Proof. Let $\ell^\infty_\sigma(I)$ denote $\ell^\infty(I)$ equipped with the topology $\sigma(\ell^\infty, \ell^1)$. Then $\ell^\infty_\sigma(I)$ is σ-compact, hence the identity $\ell^\infty_s(I) \to \ell^\infty_\sigma(I)$ is universally measurable, and it suffices to prove that the identity $\ell^\infty_\sigma(I) \to \ell^\infty_\tau(I)$ is universally measurable. This is a consequence of the following very particular fact: If K is a compact subset of $\ell^\infty_\sigma(I)$ the topology τ and σ coincide on K. Indeed if $\langle x'_i, x\rangle$ converges to $\langle x', x\rangle$, for all $x \in \ell^1(I)$ with $x'_i \in K$, the x'_i form an equicontinuous set, hence the convergence is uniform on compact subsets of $\ell^1(I)$. But in $\ell^1(I)$ weakly compact sets are compact, hence the convergence is uniform on absolutely convex $\sigma(\ell^1, \ell^\infty)$ compact subsets, i.e., x'_i tends to x' in $\ell^\infty_\tau(I)$.

The relation between universal measurability and the Orlicz property may have become apparent to the reader by now. It seems likely that a Hausdorff topology $\mathcal{T}$, on a locally convex space (or topological group) which is weaker than the given topology and such that the identity $(E, \mathcal{T}) \to E$ is universally measurable, possesses the Orlicz property. If this were so Theorem 0.2 would be a corollary of the Meyer Schwartz theorem. Whether Theorem 0.3 would follow similarly is not yet clear. We obviously have the following: let E be a Fréchet space and let $\mathscr{T}$ be a locally convex Hausdorff topology which is weaker than the given topology and such that E is essentially separable relatively to $\mathscr{T}$. Let $\vec{f} : T \to E$ be a map from a metrizable compact space T, to E which is ν-measurable relatively to $\mathscr{T}$. Then $\vec{f}$ is ν-measurable with respect to the given topology. However since here T is assumed to be metrizable the question whether the identity $(E, \mathscr{T}) \to E$ is universally measurable remains open.

§4. The Lebesgue-Nikodym theorem and absolutely summing mappings. We now again consider the general situation where

$$j : E \hookrightarrow F$$

And where $\vec{\mu} : C(T) \to E$ is an E-valued Radon mapping, and consider conditions on the injection map j which assure that if $\vec{\mu}$ is absolutely continuous with respect to ν, $j \circ \vec{\mu}$ has a density. More generally we consider not necessarily injective mappings $u : E \to F$, and study conditions under which $u \circ \vec{\mu}$ has a density for arbitrary absolutely continuous $\vec{\mu}$.

It turns out that when E and F are Banach spaces (or metrizable spaces) the relevant condition is that u be absolutely summing.

Recall that a continuous linear map $u : E \to F$ is absolutely summing when the relations $\Sigma_n |\langle x_n, x'\rangle| < +\infty$ for all $x' \in E'$ imply $\Sigma_n p(u(x_n)) < +\infty$ for all continuous semi-norms.[1] The composition on the left and on the right of an absolutely summing map with a continuous linear map furnishes an absolutely summing map. When E and F are Banach spaces, it has been shown by A. Pietsch ([1] 2. 33) that u is absolutely summing if and only if there exists a positive Radon measure λ on the $\sigma(E', E)$ compact unit ball of E' such that

$$|u(x)| \leq \int |\langle x, x'\rangle| \quad d\lambda(x')$$

for all $x \in E$. The measure λ can be chosen to be concentrated on any $\sigma(E', E)$ closed subset A for which

[1] Cf., A. Pietsch [1] 2. 1. 1.

$$|x| = \sup_{x' \in A} |\langle x, x' \rangle| \qquad \text{for all} \quad x \in E .$$

In particular a Radon map $\vec{\mu} : C(T) \to F$ is absolutely summing if and only if there exists a positive measure λ on T, such that $|\vec{\mu}(\varphi)| \leq \int |\varphi| \, d\lambda$, i. e., if and only if $\vec{\mu}$ has bounded variation, the variation being the smallest measure λ, for which the above inequality is valid.[1] If, furthermore, $\vec{\mu}$ has a density $\vec{f}$ with respect to ν, it follows that $\vec{f}$ is absolutely ν-summable, i. e., ν-integrable (cf., Proposition 2. 2).

4. 1 THEOREM. _Let_ E _and_ F _be Banach spaces and let_ $u : E \to F$ _be an absolutely summing map for which_ $|u(x)| \leq \int |\langle x, x' \rangle| \, d\lambda(x')$. _Let_ $\vec{\mu} : C(T) \to E$ _be a Radon map such that_ $\vec{\mu}$ _is weakly_ ν-_continuous_ (_or more generally such that_ $x' \circ \vec{\mu}$ _is weakly_ ν-_continuous for_ λ _almost all_ x') _and such that the map_ $x' \to x' \circ \vec{\mu} = {}^t\vec{\mu}(x')$ _is_ λ-_measurable with respect to the weak topology_ $\sigma(M, M')$. _Then_ $u \circ \vec{\mu}$ _has a density with respect to_ ν.

4. 2 COROLLARY. _If_ $\vec{\mu} : C(T) \to E$ _is a_ ν-_continuous Radon measure_ $u \circ \vec{\mu}$ _has a density with respect to_ ν.

In effect, in that case ${}^t\mu$ is even continuous relatively to the topologies $\sigma(E', E)$ and $\sigma(M, M')$.

[1] Thus the fact that a Radon map $\vec{\mu} : C(T) \to E$ is of bounded variation only depends on the topological vector space structure of $C(T)$.

4.3 COROLLARY. If λ is concentrated on a strongly separable subset of E' (in particular if E' is strongly separable) $u \circ \vec{\mu}$ has a density with respect to ν, for an arbitrary weakly ν-continuous Radon mapping $\vec{\mu}$.

In effect, in that case the identity $\sigma(E', E) \to E'_b$ (E' with norm topology) is λ-measurable by the Meyer-Schwartz theorem (or by Proposition 3.2) and ${}^t\vec{\mu} : E'_b \to M(T)$ is continuous, whence ${}^t\vec{\mu} : \sigma(E', E) \to \sigma(M, M')$ is λ-measurable.

4.4 COROLLARY. If u is a quasi-nuclear map (i.e., $|u(x)| \leq \Sigma_n \lambda_n |\langle x, x'_n \rangle|$ for all x, where $|x'_n| \leq 1$, $\Sigma|\lambda_n| < +\infty$), $u \circ \vec{\mu}$ has a density with respect to some scalar measure ν, for any Radon map $\mu : C(T) \to E$.

Indeed if we chose ν such that $x'_n \circ \vec{\mu}$ is ν-continuous for all n, Theorem 4.1 applies directly.

Remark. This last corollary is well known and obvious when u is nuclear. The proof can be reduced to this case using Corollary 2.5, and the fact established by A. Pietsch ([1] Theorem 3.2.6), that there exist an isometry $F \subsetneq G$ of F into another Banach space G, such that $\theta \circ u$ is nuclear.

We shall deduce the proof of Theorem 4.1 from the following particular case, where the stated condition is in fact both necessary and sufficient:

4.5 THEOREM. Let S be a compact space with a positive Radon measure λ, and let $u : C(S) \to L^1(S, \lambda)$ be the natural 'inclusion' map (which is injective when S = support λ). Let $\vec{\mu} : C(T) \to C(S)$ be a Radon map and let $\mu_s(\varphi) = \vec{\mu}(\varphi)(s)$. Then $\vec{\mu}$ has a density in $L^1(S, \lambda)$ (i.e., $u \circ \vec{\mu}$ has a density) with respect to ν, if and only if the following two conditions are satisfied:

a) μ_s is ν-continuous for λ-almost all s.

b) The map $s \to \mu_s$ with values in $M = C(T)'$ is λ-measurable with respect to the topology $\sigma(M, M')$ (or equivalently with respect to the norm topology).

Proof. We first remark that it follows from the theorem of Phillips-Grothendieck that if the map $s \to \mu_s$ is λ-measurable with respect to the topology $\sigma(M, M')$ this implies the measurability with respect to the norm topology.

Sufficiency. By hypothesis μ_s is ν-continuous for almost all s, hence identifying $L^1(T, \nu)$ with a closed subspace of M, $s \to \mu_s$ is a λ-measurable and bounded, almost everywhere defined map with values in $L^1(T, \nu)$, hence defines an element of $L^1(S, \lambda; L^1(T, \nu)) \simeq L^1(S, \lambda) \hat{\otimes} L^1(T, \lambda) \simeq L^1(S \times T, \lambda \otimes \nu)$. There exists then a $\lambda \otimes \mu$ integrable function $f : S \times T \to \mathbb{C}$ such that for almost all s, the function $t \to f(s, t)$ is a density of μ_s with respect to ν,

$$\mu_s(\varphi) = \int f(s, t)\, \varphi(t)\, d\nu(t) \qquad \lambda\text{-almost every } s.$$

Let $\vec{f}(t)$ be the class in $L^1(S, \lambda)$ of the function $s \to f(s, t)$. Then $\vec{f}$ belongs to $\mathscr{L}^1(T, \nu; L^1(S, \nu))$ and is the required density for $u \circ \vec{\mu}$:

Indeed, we have, with $\langle\ ,\ \rangle$ indicating duality between $L^1(S, \lambda)$ and $L^\infty(S, \lambda)$,

$$\langle u \circ \vec{\mu}(\varphi), \psi \rangle = \int \mu_s(\varphi)\, \psi(s)\, d\lambda(s)$$
$$= \iint f(s, t)\, \varphi(s)\, \psi(t)\, d\lambda(s)\, d\nu(t)$$
$$= \int \langle \vec{f}(t), \psi \rangle\, \varphi(t)\, d\nu(t)\ ,$$

whence

$$u \circ \vec{\mu}(\varphi) = \int \varphi\, \vec{f}\, d\nu\ .$$

<u>Necessity.</u> The 'inclusion' map $u : C(S) \to L^1(S, \lambda)$ is absolutely summing (since $|u(\varphi)|_{L^1} = \int |\varphi(s)|\ d\lambda(s))$; thus if $u \circ \vec{\mu}$ has a density with respect to ν, this density is absolutely summable, and corresponds to a well defined element of

$$L^1(T, \nu; L^1(S, \lambda)) \simeq L^1(T \times S, \nu \otimes \lambda)$$

We select a representative function f, so that when $\vec{f}(t)$ is the class of $f(\cdot\,, t)$ in $L^1(S, \lambda)$, $u \circ \vec{\mu} = \vec{f} \cdot \nu$. Then the function

$$s \to \int f(s, t)\, \varphi(t)\, d\nu(t)$$

is a representative of $\int \vec{f}\, \varphi\ d\nu$, hence $\mu_s(\varphi) = \int f(s, t)\, \varphi(t)\, d\nu(t)$ λ a. e. But both maps $s \to \mu_s$ and $s \to f(s, \cdot)\nu$ are measurable with respect to $\sigma(M, C)$. Hence by 2. 3 $\mu_s = f(s, \cdot)\nu$, a. e, and conditions a) and b) follow.

4.6 COROLLARY. *Let* $\vec{\mu} : C(T) \to C(S)$ *be weakly compact Radon map (Radon measure) absolutely continuous with respect to* ν. *Let* λ *be a positive Radon measure on* S. *Then there exists a* $\nu \otimes \lambda$ *-integrable function* f *such that*

$$\vec{\mu}(\varphi)(s) = \int f(s, t)\, \varphi(t)\, d\nu(t)$$

for all φ *and* s.

Proof. Since the map $s \to \mu_s$ is continuous for the topology $\sigma(M, M')$ it is λ-measurable and the above conditions are satisfied, and it follows that there exists a $\nu \otimes \lambda$ -integrable function g such that

$$\mu_s(\varphi) = \int g(s, t)\, \varphi(t)\, d\nu(t)$$

for almost all s, or more precisely that $\mu_s = g(s, \cdot)\nu$ for almost all s. But by hypothesis μ_s is ν-continuous for all s. If we select densities f_s, $\mu_s = f_s \cdot \nu$, with $f_s \in \mathscr{L}^1(\nu)$. It follows that $f_s = g_s$ in $L^1(\nu)$ for almost all s, i.e., for $s \in S_0$ with $\lambda(S - S_0) = 0$ $f(s, t) = g(s, t)$ for ν almost all t. Hence $f(s, t) = g(s, t)$ for $\lambda \otimes \nu$ almost all (s, t), whence f is $\lambda \otimes \nu$ integrable. By construction $\mu_s(\varphi) = \int f(s, t)\, \varphi(t)\, d\nu(t)$ for all s.

We now turn to the proof of Theorem 4.1 where we have

$$\vec{\mu} : C(T) \to E \qquad u : E \to F$$

$$|u(x)| \leq \int_S |\langle x, x' \rangle| \; d\lambda(x')$$

where S is the support of λ. Consider the natural map, j, from E to

$L^1(S, \lambda)$ which sends x first to the function $x' \to \langle x, x' \rangle$ on S, and then to the corresponding class of λ-equivalent functions on S. Let $\widetilde{E}$ be the closure of the image of E in $L^1(S, \lambda)$. Then it follows from the above inequality that $u(x)$ depends only on the image of x in $\widetilde{E}$, and that u can be factorized as follows:

$$u : E \xrightarrow[j]{} \widetilde{E} \xrightarrow[\dot{u}]{} F$$

where $\dot{u}$ is a continuous linear map, and $\widetilde{E}$ is a closed linear subspace of $L^1(S, \lambda)$. Since by hypothesis the map $x' \to x' \circ \vec{\mu} = \vec{\mu}_{x'}$ from S to M is λ-measurable, $\vec{\mu}_{x'}$ being ν-continuous for λ almost all x', it follows from Theorem 4.5 that $j \circ \vec{\mu}$ has a density $\vec{g}$ in $L^1(S, \lambda)$. But since $j \circ \vec{\mu}(\varphi) \in \widetilde{E}$ for all φ, it follows from Proposition 2.5 that $\vec{g}(t) \in \widetilde{E}$ for ν almost all t, and after suitable modification we can assume that $\vec{g}(t) \in \widetilde{E}$ for all t. Then since $u \circ \vec{\mu} = u \circ j \circ \vec{\mu}$, $\vec{f} = u \circ \vec{g}$ is the required density of $u \circ \vec{\mu}$ with respect to ν, by Proposition 2.8.

Thus Theorem 4.1 and its corollaries are established. Several of these corollaries had been previously obtained: 4.2 and 4.3 are due to Laurent Schwartz [2] and [3]. The case of nuclear maps had been considered by Gel'fand and others (cf., Gel'fand and Shilov [1]).

We now prove that the absolute summability of the map u is also necessary for the conclusion of Theorem 4.1 to hold, in the following sense:

4.7 THEOREM. Let E and F be quasi-complete locally convex spaces and let $u : E \to F$ be a continuous linear map. Suppose that for every Radon measure $\vec{\mu} : C[0, 1] \to E$, which is absolutely continuous with respect to Lebesgue measure ν, the measure $u \circ \vec{\mu}$ has a density with respect to ν. Then u is absolutely summing.

This is a particular case of the following more precise result.

4.7 THEOREM. Let E and F be quasi-complete locally convex spaces, and let $\mathscr{T}$ be a locally convex Hausdorff topology on F, weaker than the given topology, but such that F has a fundamental system of neighborhoods which are closed with respect to $\mathscr{T}$. We denote by $\widetilde{F}$ the space F equipped with $\mathscr{T}$. Suppose that for every Radon measure $\vec{\mu} : C[0,1] \to E$, absolutely continuous with respect to Lebesgue measure, the measure $u \circ \vec{\mu}$ has a density in $\widetilde{F}$.[1] Then $u : E \to F$ is absolutely summing (with respect to the initial topology in F).

Proof. Since we have to show $\Sigma_n \, p(u(x_n)) < +\infty$ when $\Sigma_n \, |\langle x_n , x' \rangle| < +\infty$ for all x', the problem is immediately reduced to the case where F is a Banach space, whose unit ball is closed in the topology $\mathscr{T}$, which we now suppose[2]. Let $\mathscr{L}(T, \nu; F, \mathscr{T})$ be the space of functions $\vec{f} : T \to F$ which are ν-measurable relatively to the topology $\mathscr{T}$, equipped with the topology of convergence in measure in F, for which $\vec{f}_i$ tends to zero if $\nu\{t : |\vec{f}_i(t)| \geq \varepsilon\}$ tends to zero for every positive ε. This makes

[1] Here $\widetilde{F}$ is not necessarily quasi-complete. Since we normally suppose all spaces considered to be quasi-complete it would be more correct to take $\widetilde{F}$ to be the (quasi) completion of F with respect to $\mathscr{T}$. However the hypothesis then is that the $\widetilde{F}$-valued density takes its values in F.

[2] The hypothesis implies that there is a fundamental system of semi norms of the form $p(x) = \sup_{x' \in A} |\langle x, x' \rangle|$, where A is a set of linear forms continuous with respect to $\mathscr{T}$.

sense because the norm $x \to |x|$ is lower semi-continuous with respect to $\mathcal{T}$, whence $|\vec{f}(\cdot)|$ is still ν-measurable. Let $L(T, \nu; F, \mathcal{T})$ be the set of classes of functions $\vec{f} \in \mathcal{L}(T, \nu; F, \mathcal{T})$ which are mutually equal ν-almost everywhere. We shall use the following lemma of which the proof is left to the reader.

LEMMA. $L(T, \nu; F, \mathcal{T})$ <u>is a complete metrizable topological space.</u>

(Remark: $L(T, \nu; F, \mathcal{T})$ contains $L(T, \nu; F)$ the space of classes of functions which are measurable with respect to the norm topology, as a closed subspace). Now let $x_n \in E$ and assume $\Sigma_n |\langle x_n, x' \rangle| < +\infty$ for all $x' \in E'$. Then for any sequence $(\lambda_n)_{n \in \mathbb{N}}$ tending to zero, $(\lambda_n x_n)_{n \in \mathbb{N}}$ is summable in E. Let $(f_n)_{n \in \mathbb{N}}$ be an arbitrary sequence of integrable functions on $[0, 1]$ for which $\int |f_n| \, dt \leq 1$, and put

$$\vec{\mu}_\lambda(\varphi) = \sum_n \lambda_n \int f_n \varphi \, dt \; x_n$$

Then clearly $\vec{\mu}_\lambda$ is a Radon measure which is absolutely continuous with respect to Lebesgue measure, and the measure $u \circ \vec{\mu}_\lambda$ is given by:

$$u \circ \vec{\mu}_\lambda(\varphi) = \sum_n \lambda_n \int f_n \varphi \, dt \; u(x_n) \ .$$

By hypothesis $u \circ \vec{\mu}_\lambda$ has a density $\vec{g}_\lambda$ in the space $\widetilde{F}$; $u \circ \vec{\mu}_\lambda = \vec{g}_\lambda \cdot dt$, where $\vec{g}_\lambda$ is a well defined element of $L = L([0, 1]; F, \mathcal{T})$, and the map $\lambda \to \vec{g}_\lambda$ from c_0 to L is continuous. To prove this it suffices by the previous lemma to show that its graph is closed. Now for every linear form on F, y', which is continuous with respect to $\mathcal{T}$, $y' \circ g_\lambda$ is integrable and

$$\langle u \circ \vec{u}_\lambda(\varphi), y' \rangle = \sum_n \lambda_n \int f_n \varphi \, dt \, \langle u(x_n), y' \rangle$$

$$= \int y' \circ g_\lambda \varphi \, dt ,$$

and since $\int \Sigma_n |\lambda_n| \, |f_n(t)| \, |\langle u(x_n), y' \rangle| \, dt < +\infty$ we have $y' \circ g_\lambda = \Sigma_n \lambda_n \langle u(x_n), y' \rangle f_n$ in the space $L^1[0,1]$. Thus if λ tends to zero in c_0, $y' \circ g_\lambda$ tends to zero in $L^1[0,1]$, and if by hypothesis $\vec{g}_\lambda$ converges in measure to $\vec{g}$, $y' \circ \vec{g}_\lambda$ converges in measure to $y' \circ \vec{g}$, whence $y' \circ \vec{g} = 0$ for all y', and $\vec{g} = 0$ (by 2.3), which proves that the graph is closed. Now when $(\lambda_n)_{n \in \mathbb{N}}$ is a sequence which vanishes on the complement of a finite set we obviously have

$$\vec{g}_\lambda = \sum_n \lambda_n \, u(x_n) \, f_n$$

and it follows from the previously established continuity that the same equality holds for all $\lambda \in c_0$, the series converging in $L([0,1]; F, \mathscr{T})$.

The proof now results from the following lemma:

LEMMA 2. <u>Let</u> $(y_n)_{n \in \mathbb{N}}$ <u>be a sequence in</u> F <u>such that for any sequence</u> $(f_n)_{n \in \mathbb{N}}$ <u>of functions which are integrable on</u> $[0,1]$, <u>with</u> $\int |f_n| \, dt \leq 1$, <u>the series</u> $\Sigma_n y_n f_n$ <u>converges in measure. Then</u> $\Sigma_n |y_n| < +\infty$.

Indeed if we apply this lemma to the previous situation it follows that $\Sigma |\lambda_n| \, |u(x_n)| < +\infty$ for all $(\lambda_n) \in c_0$, hence $\Sigma_n |u(x_n)| < +\infty$ which was to be proved.

Proof of Lemma 2. It follows from the hypothesis with $f_n = 1$, that y_n tends to zero, hence we may assume that $0 < |y_n| < 1$, and we suppose further that $\Sigma_n |y_n| = +\infty$. Identify the interval $[0, 1]$ with the circle of length 1 (neglecting one point), and let I_i be 'consecutive' intervals of length $|y_i|$ respectively, and put

$$f_i = \frac{1}{|y_i|} \chi_{I_i} .$$

Then $\int |f_i| = 1$ and $\Sigma_i y_i f_i$ does not converge in measure.

Indeed we can inductively define a sequence $(K_n)_{n \in \mathbb{N}}$ of disjoint (consecutive) finite subsets of $\mathbb{N}$, such that $I_i \cap I_j = \emptyset$ for $i, j \in K_n$, and

$$\sum_{i \in K_n} \nu(I_i) \geq \frac{1}{2} \qquad (\nu \text{ being Lebesgue measure}) .$$

Then

$$\left| \sum_{i \in K_n} y_i f_i(t) \right| = \sum_{i \in K_n} |y_i| f_i(t)$$

$$= \sum_{i \in K_n} \chi_{I_i}(t) \qquad \text{for} \quad 0 < \varepsilon < 1 ,$$

$$\nu\left\{ t : \left| \sum_{i \in K_n} y_i f_i(t) \right| \geq \varepsilon \right\} = \nu\left(\bigcup_{i \subset K_n} I_i \right) \geq \frac{1}{2}$$

which shows that the series does not converge in measure, and proves the theorem.

Remark. To establish Theorem 4. 7 we have only used the measurability of $\vec{f}$ relative to each seminorm. Thus the condition cannot be expected to be in general sufficient. However, with supplementary hypotheses on F, (for instance F metrizable) the absolute summability is also sufficient.

4.8 THEOREM. Let E and F be Fréchet spaces and let $u: E \to F$ be a continuous linear map. Let u be absolutely summing. Let $\vec{\mu}: C(T) \to E$ be a ν-continuous Radon measure.[1] Then $u \circ \vec{\mu}$ has a density with respect to ν. Conversely if for every Radon measure $\vec{\mu}: C[0,1] \to E$, which is absolutely continuous with respect to Lebesgue measure, ν, the measure $u \circ \vec{\mu}$ has a density, u is absolutely summing.

4.9 COROLLARY. Let E be a Fréchet space. In order that an arbitrary Radon measure with values in E, (absolutely continuous with respect to ν) have a density in E (with respect to ν) it is necessary and sufficient that E be a nuclear space.

Indeed the identity map in E is absolutely summing if and only if E is nuclear (cf., Pietsch [1] 4. 2. 5).

In particular the only Banach spaces for which the classical Lebesgue-Radon-Nikodym theorem is valid are the finite dimensional ones (see also Proposition 4. 10).

---

[1] Note that an arbitrary E valued Radon measure is ν-continuous for some scalar ν.

Proof of 4. 8. By Theorem 4. 7 it suffices to prove that the condition is sufficient, which has already been done when E and F are Banach spaces, and the present case may be reduced to that case as follows: The map $\vec{\mu} : C(T) \to E$ is weakly compact. Therefore there exists a closed, bounded, absolutely convex set $B_1 \subset E$ such that, if E_{B_1} is the Banach space with unit ball B_1, the map $\vec{\mu} : C(T) \to E_{B_1}$ is still weakly compact (cf., Grothendieck [2], Chapter IV, No. 2, Corollary 2). Since the injection $E_{B_1} \hookrightarrow E$ is continuous the map $u : E_{B_1} \to F$ is still absolutely summing, i. e.,

$$\sum_n p_k(u(x_n)) \leq M_k \sup_{|x'| \leq 1} \sum_n |\langle x_n, x' \rangle|$$

where $(p_k)_{k \geq 1}$ is a fundamental sequence of continuous seminorms in F and $|x'|$ is the norm of $x' \in E'_{B_1}$. In F let

$$|y| = \sum_{k=1}^{\infty} \frac{1}{M_k} \frac{1}{2^k} p_k(y) ,$$

then

$$\sum_n |u(x_n)| \leq \sup_{|x'| \leq 1} \sum_n |\langle x_n, x' \rangle|$$

and if $B_2 = \{y \in F : |y| \leq 1\}$, F_{B_2} is a Banach space, with $F_{B_2} \hookrightarrow F$, and $u : E_{B_1} \to F_{B_2}$ is still absolutely summing. Thus it follows from Corollary 4. 2 that $u \circ \vec{\mu}$ has a density in F_{B_2}, and a fortiori in F. Q. E. D.

In the above proof it was shown that given an absolutely summing map $u : E \to F$ from a Banach space E to a Fréchet space F, there exists a closed bounded absolutely convex set $B \subset F$ such that $u(E) \subset F_B$ and such that $u : E \to F_B$ is still absolutely summing. This is also true for a larger class of spaces F, namely those which possess property (B), defined by Pietsch ([1] 1.5-5, 1.5-8). Thus the analogue of Theorem 4.8 is also valid when E is a Fréchet space and when F is quasi-complete space with property (B). More generally it would suffice that in E every weakly compact subset A is contained in a bounded, closed, absolutely convex set B such that A is weakly compact in E_B.

Let E be a Banach space, and F a locally convex, quasi-complete space. Let us call totally summing a continuous linear map $u : E \to F$ such that there exists a bounded, closed, absolutely convex set $B \subset F$ such that $u : E \to F_B$ is absolutely summing, (in which case $u : E \to F$ is a fortiori absolutely summing). It appears from what precedes that a totally summing map transforms absolutely continuous measures into measures with densities, and conversely that any linear continuous map, u, with this property is totally summing at least when F possesses property (B). It seems likely that total summability is in all cases a necessary condition (cf., remark preceding Theorem 4.8).

We have already mentioned that the classical Lebesgue-Radon-Nikodym theorem for vector valued measures with values in a Banach space is valid only for finite dimensional Banach spaces. In the following proposition which is a sharpening of this result, it will be convenient to call a measure $\vec{\mu} : C[0,1] \to E$ absolutely continuous when it is ν-continuous where ν is Lebesgue measure.

4.10 PROPOSITION. Let E be a Banach space and suppose that for every absolutely continuous Radon measure $\vec{\mu} : C[0,1] \to E$, there exists a dense subspace $H \subset E'$ (depending on $\vec{\mu}$) and a function $\vec{f} : [0,1] \to E''$, which is a density for $\vec{\mu}$ relatively to the topology $\sigma(E'', H)$. Then E is finite dimensional.

Proof. First we prove that $\vec{f}$ is then in fact a density relatively to the topology $\sigma(E'', E')$. Indeed $\vec{f}$ is measurable relatively to this topology because E'' is σ-compact for the weak*topology (cf., § 3). If $(K_n)_{n \in \mathbb{N}}$ is a sequence of disjoint compact sets such that the restriction of $\vec{f}$ to K_n is continuous relatively to $\sigma(E'', E')$, we have

$$\int_{K_n} \vec{f}(t)\ dt = \vec{\mu}(K_n) \quad ,$$

whence the sequence $(\int_{K_n} \vec{f}\ dt)_{n \in \mathbb{N}}$ is summable in E, and a fortiori in E'' equipped with the weak*topology, which proves that $\vec{f}$ is summable in E'', (cf., Proposition 1.1). Thus the hypothesis implies that every absolutely continuous E valued Radon measure has a density in E'', relatively to $\sigma(E'', E')$. Since the unit ball of E'' is closed in the topology $\sigma(E'', E')$ it follows from Theorem 4.7 bis that the identity $E \to E''$ is absolutely summing, but this is an isometry, hence every weakly summable sequence in E is absolutely summable, whence E is finite dimensional by the theorem of Dvoretsky-Rogers.

4.11 Example. The natural injection map $\ell^1(N) \hookrightarrow \ell^2(N)$ is absolutely summing (cf., Pietsch [1], 2.4.1). Hence an arbitrary Radon measure with values in $\ell^1(\mathbb{N})$ has a density in $\ell^2(\mathbb{N})$. Thus the relations $\Sigma_n |\int \varphi f_n\, d\nu| < +\infty$ for all $\varphi \in C(T)$ imply $\Sigma_n |f_n(t)|^2 < +\infty$ ν a.e.

4.12 Example. Let $u : \mathscr{H}_1 \to \mathscr{H}_2$ be a linear map from a Hilbert space $\mathscr{H}_1$ to $\mathscr{H}_2$. Then for any Radon measure $\vec{\mu} : C(T) \to \mathscr{H}_1$, $u \circ \vec{\mu}$ has a density if and only if and only if u is a Hilbert Schmidt map.

4.13 Example. Let E be a Banach space and let $F = G'$ be the dual of a nuclear space, with $E \hookrightarrow F$. Then any E-valued weakly ν-continuous Radon map has a density in F.

A number of the presently known results about the existence of densities may be recapitulated as follows: Let $\vec{\mu} : C(T) \to E$ be a Radon measure with values in a quasi-complete locally convex space E, which is ν-continuous. Suppose $\vec{\mu}$ can be factorized as follows:

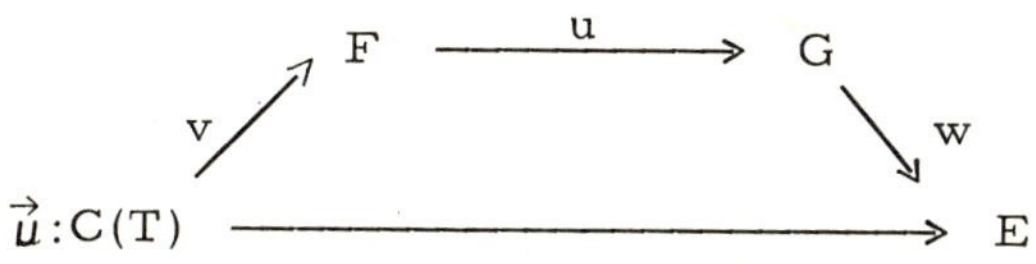

where F and G are Banach spaces and where u is absolutely summing. Then $\vec{\mu}$ has a density with respect to ν under any of the following conditions:

1) E is a Fréchet space and w is weakly compact (Grothendieck-Phillips).

2) E is the dual of a separable Banach space equipped with the weak*topology (Dunford-Pettis).

3) E is a reflexive Banach space or a strongly separable dual of a Banach space (Grothendieck-Phillips, Dunford-Pettis).

4) F is reflexive or a separable dual (L. Schwartz [2]).

5) ${}^t v$ is measurable with respect to λ , where

$|u(x)| \leq \int |\langle x, x' \rangle| \; d\lambda(x')$. In particular

6) u is quasi-nuclear.

7) v is weakly compact.

Here 1), 2) and 3) follow from the fact that $u \circ v$ being absolutely summable, hence a measure of bounded variation, extends by continuity to the space $L^1(T, |u \circ v|)$.

§5. Induced densities. Let E and F be quasi-complete locally convex spaces, with

$$E \subsetneq F$$

5.1 PROPOSITION. Let $\vec{\mu} : C(T) \to E$ be a Radon measure having an F-valued density $\vec{f}$ with respect to ν. In order that $\vec{\mu}$ have an E-valued density with respect to ν (or any other positive measure) it is necessary and sufficient that $\vec{f}(t) \in E$ a.e., and that the (almost everywhere defined) function $\vec{f} : T \to E$ be ν-measurable.

Proof. The condition is clearly necessary since any E-valued density is a fortiori an F-valued density hence by Proposition 2.6 is equal to $\vec{f}(t)$ ν-almost everywhere. A change in basis measure does not alter this because of Proposition 2.7. The condition is sufficient: To distinguish from the given $\vec{f} : T \to F$ call $\vec{g}$ a map $\vec{g} : T \to E$ equal to $\vec{f}$ almost everywhere, so that $\vec{f} = j \circ \vec{g}$ a.e. where j is the injection from E into F. It suffices to show that $\vec{g}$ is ν-summable. By hypothesis g is measurable, let $(K_n)_{n \in N}$ be sequence of disjoint compact sets such that the restriction of $\vec{g}$ to K_n is continuous. Then

$$\int_{K_n} \vec{g}\, d\nu = \int_{K_n} \vec{f}\, d\nu = \vec{\mu}(K_n) \in E$$

and $\vec{\mu}$ being countably additive the sequence

$$\left(\int_{K_n} g\, d\nu \right)_{n \in N}$$

is summable, whence $\vec{g}$ is ν-summable by Proposition 1.1 and 1.2.

5.2 THEOREM. _Let_ E _and_ F _be quasi-complete spaces with_ $E \subsetneq F$, _and suppose that_ E _satisfies any of the following conditions_

1) E _is metrizable and separable._

2) E _is metrizable and a countable union of weakly compact sets (e.g., a reflexive Banach space)._

3) E _is a countable union of compact sets (e.g., a weak* dual of a metrizable space)_[1]. _Let_ $\vec{\mu} : C(T) \to E$ _be a Radon measure having an_ F-_valued density_ $\vec{f}$ _with respect to_ ν. _Then in order that_ $\vec{\mu}$ _have an_ E-_valued density (with respect to_ ν _or any other positive measure) it is necessary and sufficient that_ $\vec{f}(t) \in E$ ν-_almost everywhere. An_ E-_valued function equal to_ $\vec{f}$ _almost everywhere then is the required density with respect to_ ν.

Proof. According to Proposition 5.1 it is sufficient to show that the function $\vec{f} : T \to E$ (after suitable modification) is ν-measurable. When E is a separable Frechet space, or a K-analytic space this follows from the theorem of Meyer-Schwartz (cf., §3). When E is a K_σ for the weak topology $\vec{f}$ is ν-measurable relatively to the weak topology and when E is furthermore metrizable $\vec{f}$ is ν-measurable by the theorem of Phillips-Grothendieck.

---

1 Or more generally a K-analytic space, i.e., a continuous image of a $K_{\sigma\delta}$, e.g., a Souslin space (cf., L. Schwartz [1]).

5.3 Example. Let $\vec{\mu}$ be a measure with values in $\ell^p(N)$; $1 \le p \le +\infty$. Then $\vec{\mu} = (\mu_n)_{n \in N}$ and if $\mu_n = f_n \cdot \nu$, $\vec{\mu}(\varphi) = \int \varphi \vec{f}\, d\nu$ with

$$\vec{f}(t) = (f_n(t))_{n \in N} ;$$

$\vec{\mu}$ has a density $\vec{f}$ with values in $\mathbb{C}^{\mathbb{N}}$. For $1 \le p \le q < +\infty$, $\vec{\mu}$ has a density in ℓ^q if and only if $\Sigma_n |f_n(t)|^q < +\infty$, ν a.e. For $q = +\infty$, $\vec{\mu}$ will have a density in ℓ^∞_τ (ℓ^∞ equipped with the topology $\tau(\ell^\infty, \ell^1)$) if and only if

$$\sup_n |f_n(t)| < +\infty, \quad \nu \text{ a.e.}$$

This follows immediately from Theorem 5.2 except for the last assertion. Here 5.2 only yields the fact that $\vec{f}$ is a density relatively to the topology $\sigma(\ell^\infty, \ell^1)$. But by 3.5 $\vec{f}$ is measurable relatively to $\tau(\ell^\infty, \ell^1)$ and hence by 5.1 $\vec{f}$ is a density relatively to $\tau(\ell^\infty, \ell^1)$. (Here, incidentally, the strengthening of the topology from $\sigma(\ell^\infty, \ell^1)$ to $\tau(\ell^\infty, \ell^1)$ has very limited interest since we have not increased the number of admissible linear forms, i.e., we have not increased the number of relations $\langle \vec{\mu}(\varphi), x' \rangle = \int \varphi \langle \vec{f}, x' \rangle \, d\nu$.) Remark that a density with values in ℓ^∞_τ is not necessarily a density in ℓ^∞, (with the norm topology). For instance if we put

$$\vec{\mu}(\varphi) = \left(\int_0^{2\pi} e^{-int} \varphi(t)\, dt \right)_{n \in \mathbb{Z}}$$

$\vec{\mu}$ is a Radon measure with values in ℓ^∞ (it is even a measure of bounded variation since $|\vec{\mu}(\varphi)| \le \int_0^{2\pi} |\varphi| \, dt$). The function $\vec{f}(t) = \{e^{-int}\}_{n \in \mathbb{Z}}$ is continuous from $[0, 2\pi]$ to ℓ^∞_τ but it is not measurable with values in ℓ^∞. This can be seen directly, but more simply by Proposition 5.1 and 2.5 which would imply $\vec{f}(t) \in c_0$ a.e., which is not the case. Use of 5.1 and 2.5 furnishes the following proposition which has some interest of its own:

5.4 PROPOSITION. *Let* ν *be a positive measure on* T, *and let* $(f_n)_{n \in N}$ *be a sequence of* ν*-integrable functions such that*

$$\lim_{n\to\infty} \int_A f_n \, d\nu = 0 \quad \text{(resp. exists)} .$$

Then $\lim_{n\to\infty} f_n(t) = 0$ a.e. (*resp.* $\lim_{n\to\infty} f_n(t)$ *exists a.e.*) *if and only if the function* $\vec{f}(t) = (f_n(t))_{n \in N}$ *has almost all values in* ℓ^∞ *and is* ν*-measurable in* ℓ^∞.

Proof. If $\lim_{n\to\infty} f_n(t)$ exists it follows from Proposition 3.2 that $\vec{f}$ is a c-valued measurable function (c the space of convergent sequences) hence an ℓ^∞-valued measurable function. Conversely if this is the case it follows from 5.1 that $\vec{f}$ is an ℓ^∞-valued density for the measure $\vec{\mu}$ defined by

$$\vec{\mu}(A) = \left(\int_A f_n \, d\nu \right)_{n \in \mathbb{N}}$$

with values in c_0 (resp. c) and hence $\vec{f}(t) \in c_0$ (resp. c) almost everywhere by Proposition 2.5.

Remark. The fact that $\vec{\mu}$ is countably additive, with values in c, results from the classical theorem of Vitali-Hahn-Saks or from the fact that the topology induced by $\mathbb{C}^{\mathbb{N}}$ in c has the Orlicz property (which is proved however with the help of the above mentioned theorem).

5.5 Example. a) Let $E = M(S)$ the space of scalar Radon measures on a compact space S, equipped with the norm topology, and let F be the same linear space equipped with the topology $\sigma(M, C)$. Then if $\vec{\mu}$ is an E-valued measure with density in F, $\vec{\mu}(\varphi) = \int \varphi \vec{f}\, d\nu$ where $\vec{f} : T \to M(S)$ is vaguely ν-summable. Since $\vec{\mu}$ is weakly compact there exists a positive measure λ on S such that $\vec{\mu}(\varphi) \in L^1(\lambda) \subset M$ for all φ. If $\vec{f}$ is measurable with respect to the norm topology it follows that $\vec{f}(t) \in L^1(\lambda)$ a.a t. Conversely if S is metrizable the relation $\vec{f}(t) \in L^1(\lambda)$ for almost all t, implies by Theorem 5.2 that $\vec{f}$ is a density relatively to the norm topology.

5.5 b) Let E be $M(S)$ equipped with the vague topology $\sigma(M, C)$ and let F be $M(S)$ equipped with some weaker topology $\sigma(M, H)$ where H is a dense subspace of C. (e.g., with $S = [a, b]$ $H = C^{\infty}[a, b]$). Then if $\vec{\mu} : C(T) \to M(S)$ has a density relatively to $\sigma(M, H)$ this is automatically a density relatively to the topology $\sigma(M, C)$. Roughly speaking: if there is a density in $M(S)$ at all it is at least vaguely measurable.

Let us now show by an example that the analogue of Proposition 5.1 or Theorem 5.2 is false if we only suppose that $\vec{\mu}$ is a Radon mapping.

5.2 Example. Let $T = [0, 1]$ let f_n be the continuous function on $[0, 1]$ which is linear in the intervals

$$\left[0, \frac{1}{n}\right], \quad \left[\frac{1}{n}, \frac{2}{n}\right], \quad \left[\frac{2}{n}, 1\right]$$

such that $f_n(0) = 0$, $f_n(1/n) = n$, $f_n(2/n) = 0$, and $f_n(1) = 0$, so that

$$\lim_{n \to \infty} f_n(t) = 0 \qquad \text{for all } t ,$$

and

$$\lim_{n \to \infty} \int_0^1 f_n \varphi \, dt = \varphi(0)$$

for all functions φ which are continuous (at 0). Put

$$\vec{\mu}(\varphi) = \left(\int_0^1 \varphi f_n \, dt \right)_{n \in N} ,$$

then $\vec{\mu}$ is a Radon map with values in c, having a density $\vec{f}$ in $\mathbb{R}^{\mathbb{N}}$, and such that $\vec{f}(t) = (f_n(t))_{n \in N} \in c$ for all t, the function $\vec{f} : [0, 1] \to c$ being measurable (by Proposition 3. 2). But $\vec{f}$ is not a c-valued density for $\vec{\mu}$, since otherwise by 2. 5 we would have $\vec{\mu}(\varphi) \in c_0$ for all φ, which is not the case. Nevertheless $\vec{f}$ is ν-measurable and $\langle \vec{f}, x' \rangle$ is ν-integrable for all $x' \in c'$, but

$$\langle \vec{\mu}(1), x' \rangle \neq \int_0^1 \langle \vec{f}, x' \rangle \, dt$$

when

$$\langle x', x \rangle = \lim_{n \to \infty} x_n .$$

<u>Remark.</u> Relations such as $\langle \vec{\mu}(\varphi), x' \rangle = \int \langle \vec{f}, x' \rangle \varphi \, d\nu$ are stipulated as definitions of the density for a Radon map. However, for Radon measures one can by the use of Proposition 5.1 obtain relations of this type which are not obvious a priori. For instance in Proposition 5.4, if we suppose

$$g(t) = \lim_{n \to \infty} f_n(t)$$

exists it follows that

$$\int_A g \, d\nu = \lim_{n \to \infty} \int_A f_n \, d\nu \ .$$

This can be proved directly using the theorem of Vitali-Hahn-Saks.

§6. Use of the variation. We shall now show that in certain cases the variation of a measure with values in a normed space indicates whether or not the measure has a density in this space (relatively to some topology which may be weaker than the topology associated with the norm).

Let F be a quasi-complete locally convex space. A function $p : F \to [0, +\infty]$ will be called a generalized semi-norm when $p(x+y) \leq p(x) + p(y)$, and $p(\lambda x) = |\lambda| p(x)$ (with $0 \times (+\infty) = 0$). We shall consider only lower semi-continuous generalized semi-norms, i.e., generalized semi-norms for which the set $B_p = \{x \in F : p(x) \leq 1\}$ is closed. Let $B_p^0 = \{x' \in F' : |\langle x, x' \rangle| \leq p(x)\}$ be the polar of B_p. Then since B_p is absolutely convex and closed the bi-polar theorem implies

$$B_p = \{x : |\langle x, x' \rangle| \leq 1 \quad \forall\, x' \in B_p^0\}$$

or

$$p(x) = \sup_{x' \in B_p^0} |\langle x, x' \rangle| .$$

Let J be a finite subset of B_p^0, and put

$$p_J(x) = \sup_{x' \in J} |\langle x, x' \rangle| .$$

Then for $J_1 \subset J_2$, $p_{J_1} \leq p_{J_2} \leq p$ and

$$p(x) = \sup_J p_J(x) .$$

Thus a lower semi-continuous generalized semi-norm is the upperbound of a filtering increasing family of (weakly) continuous semi-norms. In particular a lower semi-continuous generalized semi-norm remains lower semi-continuous if we replace the given topology by the weak topology $\sigma(F, F')$, and p can be extended to a lower semi-continuous generalized semi-norm on the weak completion of F, the algebraic dual of F', by putting

$$p(x'^*) = \sup_{x' \in B_p^0} |\langle x'^*, x' \rangle|$$

the extension still being denoted by p. In connection with measure theory we shall use the notation $p(x) = |x|_p$.

Let $\vec{\mu}: C(T) \to F$ be a Radon map. The variation of $\vec{\mu}$ with respect to p, denoted by $|\vec{\mu}|_p$, is the smallest positive measure m on the Borel σ-algebra of T for which

$$|\vec{\mu}(A)|_p \leq m(A)$$

for all Borel sets A. Explicitly

$$(6.1) \qquad \begin{aligned} |\vec{\mu}|_p(A) &= \sup_{\Sigma A_i \subset A} \sum_i |\vec{\mu}(A_i)|_p \\ &= \sup_{\substack{\Sigma A_i \subset A \\ x_i' \in B_p^0}} \sum_i \left|\vec{\mu}_{x_i'}(A_i)\right| \end{aligned}$$

where the supremum is taken over all finite sequences (A_i) of mutually disjoint Borel subsets of A.[1] It follows that $|\mu|_p$ is the smallest Borel measure m such that $|\vec{\mu}_{x'}| \leq m$ for all $x' \in B_p^0$. Clearly the

---

1 See footnote on following page.

definition is independent of the topology as long as it is compatible with the duality between F and F', i.e., as long as the dual is F'. We also have:

6.2 PROPOSITION: Let $\vec{\mu}$ be a Radon measure with values in F, and let H be a total subspace of F' such that B_p is closed for the topology $\sigma(F,H)$, (so that p is lower semi-continuous with respect to this topology). Let $\tilde{\mu}$ be the measure obtained by equipping F with the topology $\sigma(F,H)$. Then $|\tilde{\mu}|_p = |\vec{\mu}|_p$.

Proof. Since $\tilde{\mu}(A) = \vec{\mu}(A)$ this follows from Definition 6.1.

If $\vec{\mu}$ is only a Radon mapping care should be taken since in that case one might have $|\tilde{\mu}|_p \neq |\vec{\mu}|_p$.

The measure $|\vec{\mu}|_p$ is not necessarily finite, or even σ-finite, and consequently is not necessarily a Radon measure. Nevertheless we still have:

6.3 PROPOSITION:

a) $$|\vec{\mu}|_p(A) = \sup_{K \subset A} |\vec{\mu}|_p(K), \qquad K \text{ compact.}$$

b) $|\vec{\mu}|_p$ is a Radon measure if and only if the following equivalent conditions are satisfied:

i) $|\vec{\mu}|_p(T) < +\infty$

ii) There exists a positive Radon measure ν with

---

1 When $\vec{\mu}$ is a Radon map which is not a measure $\vec{\mu}(A)$ is the element of the weak completion of F defined by $\langle \vec{\mu}(A), x' \rangle = \vec{\mu}_{x'}(A)$. The set function defined by 6.1 is clearly additive, and the σ-additivity can be verified by taking a sequence A_n, with $A_n \subset A_{n+1}$ for which $|\mu|_p(A_n) \nearrow |\mu|_p(\cup A_n)$.

$$|\vec{\mu}(\varphi)|_p \leq \nu(|\varphi|) \qquad \forall\ \varphi \in C(T) \ .$$

<u>In that case</u> $|\vec{\mu}|_p$ <u>is the smallest such measure</u> ν.

<u>Proof.</u> The inner regularity follows immediately from the second expression in 6. 1. In b) i) implies ii) by taking $\nu = |\vec{\mu}|_p$ and ii) implies $|\vec{\mu}(\varphi)|_{p_J} \leq \nu(|\varphi|)$ and $|\vec{\mu}(A)|_{p_J} \leq \nu(A)$ for all finite subsets $J \subset B_p^0$; hence $|\vec{\mu}(A)|_p \leq \nu(A)$, whence $|\vec{\mu}|_p \leq \nu$.

For the next proposition we need the following lemma:

6. 4 LEMMA.

$$|\vec{\mu}|_p(A) = \sup_{J \subset B_p^0} |\vec{\mu}|_{p_J}(A)$$

<u>where</u> J <u>runs through the finite subsets of</u> B_p^0.

Since $|x|_p = \sup_J |x|_{p_J}$ for all x this follows from Definition 6. 1.

6. 5 PROPOSITION : <u>Let</u> $\vec{\mu} : C(T) \rightarrow F$ <u>be a Radon mapping</u>.

a) <u>If</u> $\vec{\mu}$ <u>is weakly</u> ν-<u>continuous</u> $\nu(A) = 0$ <u>implies</u> $|\vec{\mu}|_p(A) = 0$.

b) <u>If</u> $\vec{\mu} = \vec{f} \circ \nu$, $|\vec{\mu}|_p(A) = \int_A |\vec{f}(t)|_p \, d\nu(t)$ <u>for all Borel sets</u> A.

<u>Proof.</u> a) is obvious from the second expression in 6. 1. To prove b) first suppose that $\vec{f}$ is continuous. Then it is easy to show that

$$|\vec{\mu}|_{p_J}(A) = \int_A |\vec{f}|_{p_J} \, d\nu$$

(this also follows from the particular case of b) proved by Bourbaki [2]

Chapter 6, p. 40, Proposition 8b). The functions $|\vec{f}(\cdot)|_{p_J}$ forms a filtering increasing family of continuous functions tending to $|\vec{f}(\cdot)|_p$. Hence, by the regularity of ν it follows that

$$\int_A |\vec{f}|_p \, d\nu = \sup_J \int_A |\vec{f}|_{p_J} \, d\nu ,$$

whence $|\vec{\mu}|_p(A) = \int_A |\vec{f}|_p \, d\nu$ by the above lemma. When $\vec{f}$ is ν-measurable but no longer continuous we have $T = N + \Sigma_n K_n$ where the restriction of $\vec{f}$ to K_n is continuous and where N is a ν-negligible Borel set. Hence $|\vec{\mu}|_p(N) = 0$ and if $A_n = A \cap K_n$

$$|\vec{\mu}|_p(A) = \sum_n |\vec{\mu}|_p(A_n) = \sum_n \int_{A_n} |\vec{f}|_p \, d\nu = \int_A |\vec{f}| \, d\nu .$$

6.6 <u>Example</u>. Let $\vec{\mu} = (\mu_n)_{n \in N}$ be a measure with values in $\ell^1(N)$. Then the variation of $\vec{\mu}$ with respect to the norm is given by $|\vec{\mu}|(A) = \Sigma_n |\mu_n|(A)$. This can be seen by a direct argument but the following proof illustrates the ideas of this article: Suppose each μ_n is absolutely continuous with respect to ν. $\mu_n = f_n \cdot \nu$. Then $\vec{\mu}$ has the density $\vec{f} = (f_n)_{n \in N}$ with values in $\mathbb{C}^{\mathbb{N}}$. The function

$$x \to |x| = \sum_{n \in N} |x_n|$$

is a lower semi-continuous generalized semi-norm in $\mathbb{C}^{\mathbb{N}}$. Hence

$$|\vec{\mu}|(A) = \int_A |\vec{f}(t)|\, d\nu = \int_A \sum_n |f_n(t)|\, d\nu$$

$$= \sum_n \int_A |f_n|\, d\nu = \sum_n |\mu_n|(A) .$$

If for instance

$$T = [0, 2\pi] ,$$

$$f_n(t) = \frac{1}{n} e^{-int}$$

and ν is Lebesgue measure we have

$$|\vec{\mu}|(A) = \sum_n \frac{1}{n} \nu(A) ,$$

i.e., $|\vec{\mu}|(A) = 0$ if $\nu(A) = 0$ and $|\vec{\mu}|(A) = +\infty$ if $\nu(A) > 0$.

The next theorem shows how in certain cases the variation conversely gives information about the density.

6.7 THEOREM. Let E and F be quasi-complete locally convex spaces with $E \subsetneq F$. Let $\vec{\mu}: C(T) \to E$ be a Radon measure having an F-valued density with respect to $\nu : \vec{\mu}(\varphi) = \int \varphi \vec{f}\, d\nu$ in F. Let $x \to |x|$ be a semi-norm on E such that the set $B = \{x : |x| \leq 1\}$ is closed in F, and let $|\vec{\mu}|$ be the variation of $\vec{\mu}$ with respect to this semi-norm. Then $\vec{f}(t) \in E$ ν a.e. if and only if $|\vec{\mu}|$ is σ-finite. If furthermore B is compact in E, or if E is metrizable and B is weakly compact, $\vec{f}$ is (after suitable modification) an E-valued density for $\vec{\mu}$ with respect to ν.

Remark. Theorem 5.2 gives several other conditions on E, which guarantee that $\vec{f}$ is an E-valued density.

Proof. Note first that since the set B is closed in F it is closed in E so that $|\cdot|$ is lower semi-continuous and the variation $|\vec{\mu}|$ makes sense. Furthermore if we put $p(x) = |x|$ for $x \in E$ and $p(x) = +\infty$ for $x \in F$ and $x \notin E$, $B_p = B$ is closed in F hence p is a lower semi-continuous generalized semi-norm in F. If j is the injection $E \subsetneq F$ we have $|\vec{\mu}| = |j \circ \vec{\mu}|_p$, hence $|\vec{\mu}|(A) = \int |\vec{f}(t)|_p \, d\nu(t)$ by 6.5. If $|\vec{\mu}|(A) < +\infty$, $|\vec{f}(t)|_p < +\infty$ ν-almost everywhere on A, hence $\vec{f}(t) \in E$, ν-almost everywhere on A, and if $|\vec{\mu}|$ is σ-finite $\vec{f}(t) \in E$ ν-almost everywhere. Conversely if $\vec{f}(t) \in E$ ν-almost everywhere we may by modifying $\vec{f}$ suppose $\vec{f}(t) \in E$ for all t. Since p is lower semi-continuous on F the function $|\vec{f}(\cdot)|_p$ is finite and ν-measurable. Hence $T = N + \Sigma_n K_n$ where $\nu(N) = 0$ and the restriction of $|\vec{f}(\cdot)|_p$ to the compact K_n is continuous, hence bounded. Thus $|\vec{\mu}|(N) = 0$ and

$$|\vec{\mu}|(K_n) = \int_{K_n} |\vec{f}(t)|_p \, d\nu(t) < +\infty ,$$

whence $|\vec{\mu}|$ is σ-finite, which proves the first assertion. In the case where $\vec{f}(t) \in E$ for all t, the map $\vec{f} : T \to E$ is not necessarily ν-measurable. If however B is compact (resp. weakly compact) E is

σ-compact (resp. weakly σ-compact) and the last assertion follows from Theorem 5.2.

Remark. If E is a closed subspace of F and if $|x| = 0$ for all $x \in E$ the above theorem reduces to Proposition 2.5 (for Radon measures).

6.8 Remark. Let E be a normed space with $E \subsetneq F$ and suppose that the unit ball B of E is closed in E for the topology induced by F. Then the normed space $F_{\overline{B}}$ generated by the closure of B in F, which we shall denote by E_∞ is such that the injection $E \subsetneq E_\infty$ is an isometry. The above theorem can then be applied to E_∞ instead of E. If $\vec{\mu}$ is E-valued measure with a density $\vec{f}$ in F we have $\vec{f}(t) \in E_\infty$ a.e., if and only if $|\vec{\mu}|$ is σ-finite. Following are two examples of this.

6.9 Examples. a) The closure of the unit ball of c_0 (or c) in $\mathbb{C}^{\mathbb{N}}$ is the unit ball of ℓ^∞. Thus a c_0-valued Radon measure has a density in ℓ^∞ ($\sigma(\ell^\infty, \ell^1)$ measurable) if and only if $|\vec{\mu}|$ is σ-finite.

b) Let $\vec{\mu} : C[0,1] \to C_0^1[0,1]$ be an absolutely continuous Radon measure with values in the space of continuously differentiable functions u such that $u(0) = u(1) = 0$ with the norm

$$|u|_1 = \sup_{0 \le t \le 1} |u'(t)| .$$

The injection $C_0^1[0,1] \subsetneq C_0(0,1)$ is absolutely summing since

$$|u|_\infty \le \int_0^1 |u'(t)| \, dt$$

and the linear forms $u \to u'(t)$ form an equi-continuous set. It follows from Theorem 4.2 that $\vec{\mu}(\varphi) = \int \varphi \vec{f} \, dt$ where $\vec{f}$ is integrable with values in $C_0(0,1)$. The closure in $C_0(0,1)$ of the unit ball of $C_0^1(0,1)$ is the set of functions u such that $|u(s) - u(t)| \leq |s - t|$ and $u(0) = u(1) = 0$. It follows that $\vec{f}(t)$ is a Lipschitz function for almost all t, if and only if the variation of the measure $\vec{\mu}: C[0,1] \to C_0^1(0,1)$ is σ-finite. If for instance $\vec{\mu}(\varphi)$ is the solution of the "Dirichlet problem" $-u'' = \varphi$, $u(0) = u(1) = 0$, we have

$$|u'(t) - u'(s)| \leq \int_0^1 |\varphi| \, dt ,$$

hence by Rolle's theorem

$$\left|\vec{\mu}(\varphi)\right|_{C_0^1} = \sup_t |u'(t)| \leq \int_0^1 |\varphi| \, dt .$$

Hence $\vec{\mu}(\varphi) = \int \varphi \vec{f} \, dt$ with $\vec{f}(t)$ Lipschitz function with

$$|\vec{f}(t)(s_1) - \vec{f}(t)(s_2)| \leq |s_1 - s_2| .$$

In fact

$$\vec{f}(t) = \begin{cases} s(1 - t) & 0 \leq s \leq t \leq 1 \\ t(1 - s) & 0 \leq t \leq s \leq 1 \end{cases} .$$

Hence $\vec{f}$ is a continuous $C_0(0,1)$ valued density, such that

$$\vec{f}(t) \in \mathrm{Lip}_0(0,1) = \{u : |u(s) - u(t)| \leq M|s - t|, \ u(0) = u(1) = 0\} ,$$

(the norm of this space, $|u|_1$ being the smallest constant M, equal to ess. $\sup_t |u'(t)|$.) But since $|\vec{f}(t_1) - \vec{f}(t_2)|_1 = 1 - |t_1 - t_2|$ for

$0 < t_1 < t_2 < 1$, $\vec{f}$ is not a $\mathrm{Lip}_0(0,1)$ valued measurable function.

6.10 COROLLARY. Let $\vec{\mu} : C(T) \to E'$ be a Radon measure with values in the dual of a separable Banach space E. Let $|\vec{\mu}|$ be the variation of $\vec{\mu}$ with respect to the norm in E', and suppose $\vec{\mu}$ is ν-continuous.[1] Then $\vec{\mu}$ has a density with respect to ν if and only if $|\vec{\mu}|$ is σ-finite. If furthermore, E' is strongly separable $\vec{\mu}$ is a Radon measure $\vec{\mu}_b$ relative to the strong (norm) topology, $|\vec{\mu}_b| = |\vec{\mu}|$, and any density relatively to the topology $\sigma(E',E)$ is also a density relatively to the norm topology.

Proof. Let $(x_n)_{n\in N}$ be a dense sequence in E and use the map $x' \to \{x', x_n\}_{n\in N}$ to plunge E' into $\mathbb{C}^{\mathbb{N}}$. If $\vec{\mu}$ is ν-continuous (which is the case if each $\vec{\mu}_{x_n}$ is ν-continuous) $\vec{\mu}$ has a density $\vec{f}$ in $\mathbb{C}^{\mathbb{N}}$ with respect to ν and we are in the case described in Theorem 6.7 where furthermore B is compact in the given topology $\sigma(E',E)$ (hence compact in $\mathbb{C}^{\mathbb{N}}$), whence the first assertion. If furthermore E' is separable the topology $\sigma(E',E)$ has the Orlicz property in the normed space E', hence E' is weakly Σ-complete and $\vec{\mu} : C(T) \to E'$ is a Radon measure $\vec{\mu}_b$ with values in the normed space E'. Since $\vec{\mu}_b(A) = \vec{\mu}(A)$ we have a $|\vec{\mu}_b| = |\vec{\mu}|$, and the last assertion follows from Theorem 6.7 or 5.1

Remark.[2] Using the first part of this corollary (Dunford-Pettis theorem) and Theorem 4.7 bis we obtain the following:

---

1 Since E is separable this is always the case for some ν.

2 This remark, corollary and proof were added February 1973.

COROLLARY. Let E be an infinite dimensional Banach space. Then there exists a countably additive measure $\vec{\mu} : B[0,1] \to E$, defined on the Borel subsets of the interval $[0,1]$, absolutely continuous with respect to Lebesgue measure, and such that the variation $|\vec{\mu}|$ is not σ-finite. Such a measure $\vec{\mu}$ exists with conditionally compact range.

Proof. It suffices to prove this when E is separable. Then E may be plunged isometrically into the dual H' of a separable Banach space H. Now if every measure $\vec{\mu} : B[0,1] \to E$ has σ-finite variation it follows from 6.10 that every such measure has a density in H', so that by 4.7 bis the isometry $E \to H'$ is absolutely summing which is not the case (unless E is finite dimensional). This proves the first assertion. Inspection of the proof of 4.7 bis shows that $\vec{\mu}$ may be chosen of the form

$$\vec{\mu}(A) = \sum_n \int_A f_n \, dt \, x_n$$

where $(x_n)_{n \in N}$ is summable but not absolutely summable, and $f_n \in \mathscr{L}^1[0,1]$ with

$$\int |f_n| \, dt \leq 1 \quad \text{and} \quad \lim_n \int |f_n| \, dt = 0$$

Thus the range of $\vec{\mu}$ is contained in the set

$$\left\{ \sum_n \alpha_n x_n \;\middle|\; |\alpha_n| \leq 1 \right\}$$

which is compact.

6.11 <u>A counterexample</u>. If E is a non-separable Banach space and $\vec{\mu}: C(T) \to E'$ is a measure of bounded variation there does not necessarily exist an E' valued density (relative to $\sigma(E', E)$) even if $T = [0, 1]$ and consequently $\vec{\mu}$ has separable range. Take for instance $E = \ell^1(I)$, whence $E' = \ell^\infty(I)$. Let $(u_i)_{i \in I}$ be a family of elements of $L^\infty[0, 1]$, with $\|u_i\|_\infty \leq 1$. Put

$$\vec{\mu}(\varphi) = \left(\int \varphi u_i \cdot dt \right)_{i \in I} .$$

Then $\vec{\mu}: C[0, 1] \to \ell^\infty(I)$ and $|\vec{\mu}(\varphi)| \leq \int |\varphi| \, dt$. If $\vec{\mu}(\varphi) = \int \varphi \vec{f} \, dt$ where $\vec{f} : [0, 1] \to E'$ is a density relatively to $\sigma(E', E)$ we have $f_i \in u_i$ and there exists a partition $[0, 1] = N + \Sigma_n K_n$ such that $f_i|_{K_n}$ is continuous for every i. If

$$I = \{ u \in L^\infty[0, 1] : \|u\|_\infty \leq 1 \}$$

and $i \to u_i$ is the identity mapping this is clearly impossible.

Of course there always exist functions $\vec{f} : [0, 1] \to \ell^\infty(I) = E'$ such that for any $x \in E = \ell^1(I)$, $\langle \vec{f}, x \rangle$ is integrable and

$$\langle \vec{\mu}(\varphi), x \rangle = \int \langle \vec{f}, x \rangle \varphi \, dt \qquad \text{for all } \varphi .$$

In the above example it suffices to take $\vec{f} = (f_i)_{i \in I}$ with any choice $f_i \in u_i$. However, such functions are not essentially unique. For instance if $I = [0,1]$ and $g_i(t) = f_i(t)$ for $t \neq i$ and $g_i(t) \neq f_i(t)$ for $t = i$, one also has

$$\langle \vec{\mu}(\varphi), x \rangle = \int \langle \vec{g}, x \rangle \varphi \, dt \quad \text{for all } x \in E \text{ and all } \varphi \in C[0,1],$$

but $\vec{g}(t) \neq \vec{f}(t)$ for all t.

On the other hand since $C[0,1]$ is separable $\vec{\mu}$ has separable range. Suppose therefore that $\vec{\mu}: C[0,1] \to G \subset E'$ where G is a strongly separable strongly closed subspace of E', and $|\vec{\mu}(\varphi)| \leq \int |\varphi| \, dt$ as before. Then there exists a separable subspace H of E such that the restriction map $E' \to H'$ reduces to an isometry on G,[1] i.e., $G \subsetneq H'$, G being a strongly closed subspace of H'. Consequently $\vec{\mu}$ being of bounded variation, we have $\vec{\mu}(\varphi) = \int \varphi \vec{f} \, dt$ where $\vec{f} : [0,1] \to H'$ is a density relatively to the topology $\sigma(H', H)$. However since G is not necessarily closed in H' for the topology $\sigma(H', H)$, we do not necessarily have $\vec{f}(t) \in G$ a.e. In fact from Theorems 2.5 and 5.2 (or 3.2) it follows that $\vec{f}(t) \in G$ a.e., if and only if $\vec{\mu}$ has a strongly measurable E'-valued density. Since by the above example there is not even in general a $\sigma(E', E)$ -measurable density it follows that we do not have $\vec{f}(t) \in G$ a.e. in general.[2]

The following theorem is a corollary of 6.10 and of the theorem of Grothendieck-Phillips.

6.12 THEOREM. Let $\vec{\mu}: C(T) \to E$ be a ν-continuous Radon measure with values in a reflexive Banach space. Then $\vec{\mu}$ has a density with respect to ν if and only if $|\vec{\mu}|$ is σ-finite.

Proof. When E is separable this is a particular case of 6.10. Therefore we need only show that both the hypothesis of the existence of a

1. Cf., Dunford and Schwartz [1] VI. 8. 7

2. If it were possible to conclude to $f(t) \in G$ a.e., any measure of bounded variation with values in a separable Banach space G would have a density since $G \subset G''$.

density, and the σ-finiteness of $|\vec{\mu}|$ imply that the image of $\vec{\mu}$ is separable. When $\vec{\mu} = \vec{f} \cdot \nu$ this is clearly the case since if $\vec{f}|_{K_n}$ is continuous $\vec{f}(K_n)$ is metrizable and compact, hence separable. If $|\vec{\mu}|(A) < +\infty$, $\vec{\mu}$ induces a map

$$\vec{\mu}|_A : L^1(A, |\vec{\mu}|) \to E$$

which is weakly compact since E is reflexive, and hence has a separable image. Hence if $|\vec{\mu}|$ is σ-finite there exists a separable closed subspace $E_0 \subset E$ such that $\vec{\mu}(A) \in E_0$ for all A, hence $\vec{\mu} : C \to E_0$.

The following result generalizes 6.10.

6.13 THEOREM. _Let_ $\vec{\mu} : C(T) \to E$ _be a_ ν-_continuous Radon measure_. _Let_ $x \to |x|$ _be a semi-norm in_ E _such that_ $B = \{x : |x| \leq 1\}$ _is closed, and let_ $|\vec{\mu}|$ _be the variation of_ $\vec{\mu}$ _with respect to this semi-norm_. _In order that_ $\vec{\mu}$ _have a density with respect to_ ν _it is necessary that_ $|\vec{\mu}|$ _be_ σ-_finite_. _If_ B _is compact and metrizable this condition is also sufficient_.

Proof. Since $|\vec{\mu}|(A) = \int_A |\vec{f}| \, d\nu$ the necessity is obvious. Suppose B is metrizable and compact. Then B may be plunged into $\mathbb{C}^{\mathbb{N}}$, and we have the injection $E_B \subsetneq \mathbb{C}^{\mathbb{N}}$. Now by the regularity of $|\vec{\mu}|$, $T = N + \Sigma_n K_n$ where K_n is compact, N ν-negligible and $|\vec{\mu}|(K_n) < +\infty$. Then the restriction of $\vec{\mu}$ to K_n is an E_B-valued Radon measure, which by Theorem 6.7 has a density f_n, with values in $\lambda_n B$, with respect to $|\vec{\mu}|$ where $\lambda_n = |\vec{\mu}|(K_n)$. The function defined almost everywhere in T by putting $f(t) = \rho_n f_n(t)$ for $t \in K_n$ where $|\vec{\mu}|_{K_n} = \rho_n \nu|_{K_n}$ is an

E-valued ν-measurable function whose restriction to each K_n is ν-summable and such that $\vec{\mu}(A) = \int_A \vec{f}\, d\nu$ for each $A \subset K_n$. It follows that $\vec{f}$ is a density relatively to ν.

Remark. Example 6.11 shows that the conclusion does not necessarily hold if B is not metrizable.

§7 Continuous densities. Let T be a compact space and let ν be a positive Radon measure on T which is dense (support $\nu = T$). Then if $\vec{\mu}: C(T) \to E$ is a Radon measure $\vec{\mu}$ has at most one density $\vec{f}$ with respect to ν which is a continuous function $\vec{f}: T \to E$ (relative to the given topology on E or any weaker locally convex Hausdorff topology). In what follows we give some conditions insuring the existence of continuous, or weakly continuous, densities.

7.1 PROPOSITION. Let $\vec{f}: T \to E$ be a continuous function from T to a quasi-complete locally convex space E. Let M_1 be the set of scalar Radon measures on T, ρ, such that

$$\|\rho\| = \sup_{|\varphi| \leq 1} |\langle \rho, \varphi \rangle| \leq 1 ,$$

equipped with the weak* topology $\sigma(M_1, C)$. Then the map $\rho \to \int \vec{f}\, d\rho$ from M_1 to E is continuous.

Proof. It suffices to prove this when E is a Banach space which we suppose. Let $\rho_i \subset M_1$ and suppose $\langle \rho_i, \varphi \rangle$ tends to $\langle \rho, \varphi \rangle$ for each $\varphi \in C(T)$. Then $\langle \int \vec{f}\, d\rho_i, x' \rangle = \int \langle \vec{f}, x' \rangle\, d\rho_i$ tends to $\int \langle \vec{f}, x' \rangle\, d\rho = \langle \int \vec{f}\, d\rho, x' \rangle$ for all $x' \in E'$. However, since the ρ_i are bounded, hence equicontinuous, $\langle \rho_i, \varphi \rangle$ tends to $\langle \rho, \varphi \rangle$ uniformly on compact sets of φ. Now if $\vec{f}$ is continuous the functions $\langle \vec{f}, x' \rangle$ form a bounded and equicontinuous, hence relatively compact subset of $C(T)$, when $|x'| \leq 1$, and it follows that the above integrals converge uniformly with respect to x', for $|x'| \leq 1$, whence $|\int \vec{f}\, d\rho_i - \int \vec{f}\, d\rho|$ tends to zero.

In what follows we shall denote by $C_1(\nu)$ the set of continuous functions φ on T, for which $\int|\varphi|\, d\nu \leq 1$, identified to a subset of M_1, equipped with the topology induced by $\sigma(M, C)$ (φ_i tends to φ if $\int \varphi_i \psi\, d\nu$ tends to $\int \varphi\psi\, d\nu$ for all continuous ψ).

7.2 PROPOSITION. Let ν be a dense measure on T and let $\vec{\mu}: C(T) \to E$ be a Radon map with values in a quasi-complete locally convex space E. Then $\vec{\mu}$ has a continuous density with respect to ν if and only if $\vec{\mu}: C_1(\nu) \to E$ is continuous with respect to the topology induced by $\sigma(M, C)$.

Proof. The necessity follows from the previous proposition. To prove the sufficiency we show that $\vec{\mu}$ is uniformly continuous on $C_1(\nu)$. Let W be a zero neighborhood in E and let V be a zero neighborhood in $\sigma(M_1, C)$ such that for $\varphi \in C_1(\nu)$ $\varphi \in V$, $\vec{\mu}(\varphi) \in W$. Then for $\varphi, \psi \in C_1(\nu)$ and

$$\frac{\varphi - \psi}{2} \in V \qquad \vec{\mu}\left(\frac{\varphi - \psi}{2}\right) \in W,$$

hence for $\varphi, \psi \in C_1(\nu)$ $\varphi - \psi \in 2V$, $\vec{\mu}(\psi) \in 2W$, which proves that $\vec{\mu}$ is uniformly continuous on $C_1(\nu)$, hence extends continuously to the closure of $C_1(\nu)$ in M, which by the bipolar theorem is M_1. If $\vec{f}(t)$ is the value at $\delta_{(t)}$ of the continuous extension, $\vec{f}$ is a continuous function on T, and the continuous extension is the map $\rho \to \vec{f}\, d\rho$. In particular, $\int \vec{\mu}(\varphi) = \int \vec{f} \varphi\, d\nu$. Q. E. D.

Let $\vec{\mu} : C(T) \to E$ be a Radon measure which is continuous on $C_1(\nu)$ for the topologies $\sigma(M, C)$ and $\sigma(E, E')$. By the previous proposition this is the necessary and sufficient condition for $\vec{\mu}$ to have a continuous density with values in E'^*, the weak completion of E. However, since $\vec{f}(t) = \lim_i \vec{\mu}(\varphi_i)$ for a family φ_i with $\varphi_i \geq 1$ and $\int \varphi_i \, d\nu = 1$ (an approximation of $\delta_{(t)}$), it follows that $\vec{f}(t)$ is in the closure of a bounded subset of E, in the space E'^* equipped with the topology $\sigma(E'^*, E')$, which implies that $\vec{f}(t) \in E''$. Hence we have:

7.3 PROPOSITION. Let $\vec{\mu} : C(T) \to E$ be a Radon map. Then $\vec{\mu}$ has a continuous density with respect to ν, in the space E'' equipped with the topology $\sigma(E'', E')$ if and only if $\vec{\mu}$ is continuous on $C_1(\nu)$ relatively to the topology $\sigma(E, E')$.

This implies the following proposition:

7.4 PROPOSITION. Let T be a metrizable compact space and suppose that E is weakly sequentially complete. Then $\vec{\mu}$ has a weakly continuous density in E, with respect to ν, if and only if $\vec{\mu}$ is continuous on $C_1(\nu)$ relatively to the topologies $\sigma(M, C)$ and $\sigma(E, E')$.

Proof. For every $t \in T$ we can find a sequence $\varphi_n \in C(T)$ such that $\varphi_n \geq 0$ $\int \varphi_n \, d\nu = 1$ supp φ_n tends to $\{t\}$. Then $\varphi_n \cdot \nu$ tends to $\delta_{(t)}$ and

$$\vec{f}(t) = \lim_{n \to \infty} \vec{\mu}(\varphi_n)$$

in the weak topology, whence $\vec{f}(t) \in E$.

The example at the end of this section shows that the conclusion does not necessarily hold if E is not weakly sequentially complete.

7.5 PROPOSITION. Let $\vec{\mu} : C(T) \to E'$ be a Radon measure with values in the dual of some Banach space (or more generally a barrelled space). Then $\vec{\mu}$ has a $\sigma(E', E)$ continuous density with respect to ν if and only if $\vec{\mu}$ is continuous on $C_1(\nu)$ with respect to the topology $\sigma(E', E)$.

Proof. E', equipped with the topology $\sigma(E', E)$, is quasi-complete, and the result follows from 7.2.

We now give an example where $T = [0, 1]$, ν is Lebesgue measure, $\vec{\mu} : C[0, 1] \to E$ is a Radon measure with values in a separable Banach space E, having an E-valued density with respect to ν, but having a unique continuous density in E'', with respect to $\sigma(E'', E')$, which does not take all its values in E.

7.6 Example. $E = c_0$, $E' = \ell^1$, $E'' = \ell^\infty$. Let $(f_n)_{n \in \mathbb{N}}$ be a sequence of continuous functions on $[0, 1]$ such that

$$|f_n| \le 1, \quad \lim_{n \to \infty} f_n(t) = 0$$

for $0 \le t < 1$, and $f_n(1) = 1$ (e.g., $f_n(t) = t^n$). Put

$$\vec{\mu}(\varphi) = \left\{ \int \varphi f_n \, dt \right\}_{n \in \mathbb{N}} \in c_0 .$$

Then $\vec{\mu}$ is a c_0-valued Radon measure (of bounded variation). Put

$$\vec{f}(t) = (f_n(t))_{n \in \mathbb{N}}$$

Then $\vec{f}$ is a continuous density in ℓ^∞, for the topology $\sigma(\ell^\infty, \ell^1)$, and since $\vec{f}(t) \in c_0$ except for $t = 1$, $\vec{\mu}$ has a density in c_0 (but no weakly continuous density in c_0).

We could restrict the space T to $[0, 1)$, obtaining a weakly continuous density with values in c_0, but this is artificial. One could complicate the example and have a dense subset of exceptional points.

This example shows that even for a measure of bounded variation, having a density in the given space, it may be of interest to consider densities "going out of the space".

REFERENCES

[1] Bartle, R.G., Dunford, N. and Schwartz, J. [1], Weak compactness and vector measures, Canadian J. Math., 7 (1955), pp. 289-305. MR16, 1123.

[2] Bessaga, C. and Pełczynski, A.A. [1], On bases and unconditional convergence of series in Banach spaces, Studia Math., 17 (1958), pp. 151-164. MR22 #5872.

[3] Bourbaki, N., Espaces vectoriels topologiques, Hermann, Paris.

[4] ________, Intégration Chapters 1 to 5, Chapter 5, 2nd Edition, Chapter 9. Paris, Hermann, 1956, 1965.

[5] Dieudonné, J., Sur le théoreme de Lebesgue-Nikodym V, Canadian J. Math., 3 (1951), pp. 129-139. MR13, 448.

[6] Dinculeanu, N. [1], Vector measures, (Pergamon Press).

[7] Dunford, N. and Pettis, B.J. [1], Linear operations on summable functions, Trans. Amer. Math. Soc., 47 (1940), pp. 323-392. MR1, 338.

[8] Gel'fand, I.M. and Shilov, G.E. [1], Generalized functions, Academic Press.

[9] Grothendieck, A. [1], Sur les applications lineares faiblement compactes d'espaces du type C(K), Canadian J. Math., 5 (1953), pp. 129-173. MR15, 438.

[10] ________, [2] Topological vector spaces, Gordon and Breach (to appear) or Espaces vectoriels topologiques (São Paulo 1964).

[11] Métivier, M. [1], Martingales a valeurs vectorielles, Ann. Inst. Fourier, Grenoble 17, 2 (1967), pp. 175-208.

[12] Pełczynski, A.A. [1], Projections in certain Banach spaces, Studia Math., 19 (1960), pp. 209-228. MR23 #A3441.

[13] Pettis, B.J. [1], On integration in vector spaces, Trans. Amer. Math. Soc., 44 (1938), pp. 277-304.

[14] Pietsch, A. [1], Nukleare Lokalkonvexe Räume, Berlin, Akademie-Verlag 1965.

[15] Phillips, R.S. [1], On linear transformations, Trans. Amer. Math. Soc., 48 (1940), pp. 515-541. MR2, 318.

[16] Rieffel, M.A. [1], The Radon Nikodym theorem for the Bochner integral.

[17] __________ [2], Dentable subsets of Banach spaces with applications to a Radon-Nikodym theorem, to appear in Trans. Amer. Math. Soc.

[18] Schwartz, L. [1], Lectures on the theory of Radon measures on arbitrary topological spaces, Tata Institute of Fundamental Research, Bombay, India.

[19] __________ [2], Seminaire Laurent Schwartz 1969/70, Centre de Mathématique, Ecole Polytechnique, Paris.

[20] __________ [3], Radon measures in arbitrary topological spaces, Tata Institute of Fundamental Research Publication, Bombay, India, to appear.

[21] Thomas, G. Erik [1], L'intégration par rapport à une mesure de Radon vectorielle, Ann. Institut Fourier, Vol. XX, No. 2 (1970) Séminaire Choquet: 1967/68, Initiation à l'analyse, Fasc. 2: Analyse Fonctionnelle Générale, Exp. B.10, 32 pp. Secrétariat mathématique, Paris, 1969.

Yale University
New Haven, Connecticut